AF676826

Integrated Control of Stored Products Pests & Diseases

Integrated Control of Stored Products Pests & Diseases

Editor

Surendra Mohite

scitus
academics

Integrated Control of Stored Products Pests & Diseases

Edited by **Surendra Mohite**

ISBN: 978-1-68117-071-8

Library of Congress Control Number: 2015935469

© 2016 by
SCITUS Academics LLC,
www.scitusacademics.com
Box No. 4766, 616 Corporate Way,
Suite 2, Valley Cottage,
NY 10989

This book contains information obtained from highly regarded resources. Copyright for individual articles remains with the authors as indicated. All chapters are distributed under the terms of the Creative Commons Attribution License, which permits unrestricted use, distribution, and reproduction in any medium, provided the original author and source are credited.

Notice

Reasonable efforts have been made to publish reliable data and views articulated in the chapters are those of the individual contributors, and not necessarily those of the editors or publishers. Editors or publishers are not responsible for the accuracy of the information in the published chapters or consequences of their use. The publisher believes no responsibility for any damage or grievance to the persons or property arising out of the use of any materials, instructions, methods or thoughts in the book. The editors and the publisher have attempted to trace the copyright holders of all material reproduced in this publication and apologize to copyright holders if permission has not been obtained. If any copyright holder has not been acknowledged, please write to us so we may rectify.

Printed in United States of America on Acid Free Paper ♾

Contents

Preface

Biology and ecology of organisms for biological control, and various facets of their use including biological contributions to integrated pest management (IPM), such as plant resistance, pheromones and intercropping. Interdisciplinary identifications with a global perspective on the use of biological control in IPM systems are strongly encouraged. Developments in molecular biology that have direct relevance to biological control are also considered for publication. Pest Technology provides authoritative coverage of research developments and advances in the fight against pests that affect plants or plant production systems. Through peer-reviewed original research papers and informative contributions on current topics, the journal acts as a bridge between academic research and application. We focus on plant protection and damage inflicted by insects, fungi, bacteria, weeds, invertebrates and vertebrates. Pest Technology reports in the areas of agriculture, horticulture, forestry, conservation, stored products research, and health and safety aspects.

Editor

Chapter 1

Botanicals as Grain Protectants

Yallappa Rajashekar[1], Nandagopal Bakthavatsalam[1], and Thimmappa Shivanandappa[2]

[1]National Bureau of Agriculturally Important Insects, P. Bag No:2491, H.A. Farm Post, Bellary Road, Karnataka, Bangalore 560 024, India

[2]Department of Zoology, University of Mysore, Manasagangotri, Karnataka, Mysore 560007, India

ABSTRACT

Prevention of food losses during postharvest storage is of paramount economic importance. Integrated pest management is now a widely accepted strategy in pest control including postharvest infestation control which involves the use of chemical (contact/residual) insecticides along with fumigants. The use of synthetic chemical insecticides is either not permitted or used restrictively because of the residue problem and health risks to consumers. In view of the above, there is a need for plants that may provide potential alternatives to the currently used insect control agents as they constitute a rich source of bioactive molecules. Available literature indicates that plant could be source for new insecticides. Therefore, there is a great potential for a plant-derived insecticidal compounds. This paper focuses on the current state of the botanical insecticides as grain protectants and its mode of action.

INTRODUCTION

Food grain losses due to insect infestation during storage are a serious problem, particularly in the developing countries [1, 2]. Losses caused by insects include not only the direct consumption of kernels, but also accumulation of exuviae, webbing, and cadavers. High levels of the insect detritus may result in grain that is unfit for human consumption and loss of the food commodities, both, in terms of quality and quantity. Insect infestation-induced changes in the storage environment may cause warm moist "hotspots" that provide suitable conditions for storage fungi that cause further losses. It is estimated that more than 20,000 species of field and storage pests destroy approximately one-third of the world's food production, valued annually at more than $100 billion among which the highest losses (43%) occurring in the developing world [3, 4]. The quantitative and qualitative damage to stored grains and grain product from the insect pests may amount to 20–30% in the tropical zone and 5–10% in the temperate zone [5, 6]. Food grain production in India has reached 250 million tonnes in the year 2010-2011, in which nearly 20–25% food grains are damaged by stored grain insect pests [7, 8]. The efficient control and removal of stored grain pests from food commodities has long been the goal of entomologists throughout the world.

The major pests of stored grain and pulses of the Indian subcontinent are classified in to two groups, namely, primary pests: those which are capable of penetrating and infesting intact kernel of grain and have immature stages develop within kernel of grain and secondary pests which cannot infest the whole grain but feed on as broken kernels, debris, high moisture weed seeds, and grain damaged by primary pests. In general, the immature stages of the secondary pest species are found external to the grain. It is often thought that secondary invaders cannot initiate infestation. The important primary pests are the rice weevil, Sitophilus oryzae (L.), granary weevil, Sitophilus granaries (L.), (Coleoptera: Curculionidae), lesser grain borer,Rhyzopertha dominica (F.), (Coleoptera: Bostrichidae), Khapra beetle, Trogoderma granarium (Everts), (Coleoptera: Dermestidae), and the pulse beetle Callosobruchus chinensis (L.) (Coleoptera: Bruchidae). The secondary pests are rust-red flour beetle, Tribolium castaneum (Herbst), (Coleoptera: Tenebrionidae), rusty grain beetle, Cryptolestes ferrugineus (L.), (Coleoptera: Cucujidae), sawtoothed grain beetle, Oryzaephilus surinamensis (L.), (Coleoptera:

Silvanidae), mites, (Acarina: Tetranychidae) Liposcelis corrodens, and (Psocoptera: Liposcelidae).

INFESTATION CONTROL BY PESTICIDES AND THEIR SIDE EFFECTS

Since the 1950s, synthetic insecticides have been used extensively in grain facilities to control stored product insect pests. Fumigants such as methyl bromide, phosphine, cyanogens, ethyl formate, or sulfuryl fluoride rapidly kill all life stages of stored product insects in a commodity or in a storage structure. Fumigation is still one of the most effective methods for the prevention of stored product losses from insect pests. But pests develop resistance, not stored products were showing a slow upsurge in fumigation resistance [9]. Resistance to phosphine is so high in Australia and India, it may cause control failures [10, 11]. Methyl bromide has been identified as a major contributor to ozone depletion [12] and has been banned in developed countries, and developing nations have committed to reducing the use by 20% in 2005 and phase out in 2015. Contact insecticides such as malathion, chlorpyrifos, or deltamethrin are sprayed directly on grain or storage structure for protection from infestation for several months. The incidence of insecticide resistance is a growing problem in stored-product protection. Resistance to one or more insecticides has been reported in at least 500 species of insects and mites [13].

Champ [14] reported that resistance to pesticides used to protect grain and other stored food stuffs is widespread and involves all groups of pesticides and most of the important pests. Some of the contact insecticides have become ineffective because of widespread resistance in insect population. Resistance to malathion is widespread in Canada, USA and Australia [15]. Stored product insects pests were found to be resistant against different insecticides including the cyclodienes, chlorpyrifos, cyanophos, carbamates, carbaryl, cypermethrin, dichlorodiphenyltrichloroethane, deltamethrin, diazinon, dichlorovos, ethylene bromide, ethyl formate organophosphates, permethrin, pyrethrins, and propoxur.

Although chemical insecticides are effective, their repeated use has led to residual toxicity, environmental pollution and an adverse effect on food besides side effect on humans [16, 17]. Their uninterrupted and indiscriminate use not only has led to the development of resistant strains but also accumulation of toxic residues on food grains used for human consumption that has led to the health hazards [18]. In view of all these problems, several insecticides have either been banned or restricted in their use.

BOTANICALS AS ALTERNATIVE TO SYNTHETIC PESTICIDES

The increasing serious problems of resistance and residue to pesticides and contamination of the biosphere associated with large-scale use of broad spectrum synthetic pesticides have led to the need for effective biodegradable pesticides with greater selectivity. This awareness has created a worldwide interest in the development of alternative strategies, including the discovery of newer insecticides [19, 20]. However, newer insecticides will have to meet entirely different standards. They must be pest specific, nonphytotoxic, nontoxic to mammals, ecofriendly, less prone to pesticide resistance, relatively less expensive, and locally available [21]. This has led to re-examination of the century-old practices of protecting stored products using plant-derivatives, which have been known to resist insect attack [5, 22–24]. Plant derived materials are more readily biodegradable, less likely to contaminate the environment and nay be less toxic to mammals. There are many examples of very toxic plant compounds. Therefore, today, researchers are seeking new classes of naturally occurring insecticides that might be compatible with newer pest control approaches [2, 25, 26].

Since ancient times, there have been efforts to protect harvest production against pests. The Egyptian and Indian farmers used to mix the stored with fire ashes [83, 84]. The ancient Romans used false hellebore (Veratrum album) as a rodenticide, the Chinese is credited with discovering the insecticidal properties of Derris species, whereas pyrethrum was used as an insecticide in Persia and China [4]. In many parts of the world, locally available plants are currently in wide use to protect stored products against damage caused by insect infestation [80, 85–87]. Indian farmers used neem leaves and seed for the control of stored grain pests [88]. In northern Cameroon, cowpeas are traditionally mixed with sieved ash after threshing and the mixture put into mud granaries or clay jars [89]. In eastern Africa, leaves of the wild shrub Ocimum suave and the cloves of Eugenia aromatic are traditionally used as stored grain protectants [90]. In Rwanda, farmers store edible beans in a traditional closed structure (imboho) and whole leaves of Ocimum canum are usually added to the stored foodstuff to prevent insect damage within these structures [75]. Owusu [91] suggested natural and cheaper methods for the control of stored product pests of cereals, with traditionally useful Ghanaian plant materials. In some south Asian countries, food grains such as rice or wheat are traditionally stored by mixing with 2% turmeric powder [92, 93]. The use of oils in stored-products pest control is also an ancient practice. Botanical insecticides such as pyrethrum, derris, nicotine, oil of citronella, and other plant extracts have been used for centuries [27, 94, 95]. More than 150 species of forest and roadside trees in India produce oilseeds, which have been mainly used for lighting, medicinal purposes, and also as insecticides from ancient times to early 20th century [96]. Turmeric, garlic, Vitex negundo, gliricidia, castor, Aristolochia, ginger, Agave americana, custard apple, Datura, Calotropis, Ipomoea, and coriander are some of the other widely used botanicals to control and repel crop pests [81, 97].

Talukder [5] has listed 43 plant species as insect repellents, 21 plants as insect feeding deterrents, 47 plants as insect toxicants, 37 plants as grain protectants, 27 plants as insect reproduction inhibitors, and 7 plants as insect growth and development inhibitors. Eighteen species showed insecticidal potential, and antiovipositional properties against Sitophilus oryzae [98].

CLASSIFICATION OF BOTANICAL INSECTICIDES

On the basis of physiological activities on insects, Jacobson [3] conventionally classified the plant components into 6 groups, namely, repellents, feeding deterrents/antifeedants, toxicants, growth retardants, chemosterilants, and attractants. Focus on the toxicants and grain protectants on activity of essential oil, extracts, and its constituents has sharpened since the 1980s.

Repellents

The repellents are desirable chemicals as they offer protection with minimal impact on the ecosystem, as they drive away the insect pest from the treated materials by stimulating olfactory or other receptors. Repellents from plant origins are considered safe in pest control; minimise pesticide residue; ensure safety of the people, food, and environment [1, 5, 99]. The plant extracts, powders, and essential oil from the different bioactive plants were reported as repellent against stored grain insect pests [1, 91, 100–102]. For example, the essential oil of Artemisia annua was found as repellent against Tribolium castaneum and Callosobruchus maculates[103].

Antifeedants/Feeding Deterrents

Antifeedants, sometimes referred to as "feeding deterrents" are defined as chemicals that inhibit feeding or disrupt insect feeding by rendering the treated materials unattractive or unpalatable [104, 105]. Some naturally occurring antifeedants, which have been characterized, include glycosides of steroidal alkaloids, aromatic steroids, hydroxylated steroid meliantriol, triterpene hemiacetal, and others [3, 106]. Essential oil constituents such as thymol, citronellal and α-terpineol are effective as feeding deterrent against tobacco cutworm,Spodoptera litura synergism, or additive effects of combination of monoterpenoids from essential oils have been reported against Spodoptera litura larvae [107]. The screening of several medicinal herbs showed that root bark of Dictamnus dasycarpus possessed significant feeding deterrence against two stored-product insects [108].

Toxicants

Research on new toxicants of plant origin has not declined in recent years despite the increased research devoted to the discovery of synthetic insecticides [25]. Worldwide reports on plant derivates showed that many plant products are toxic to stored product insects [6, 16, 27, 55, 82, 91, 109–114]. Talukder [32] listed the use of 43 plant species expressing toxicant effects of different species of stored-products insects. Pascual-Villalobos and Robledo [115] carried out screening of plant extracts from 50 different wild plant species of southeastern Spain for insecticidal activity towards Tribolium castaneum and reported that four species, namely, Anabasis hispanica, Senecio lopezii, Bellardia trixago, and Asphodelus fistulosus were found be promising. Two major constituents of the essential oil of garlic, Allium sativum, methyl allyl disulfide and diallyl trisulfide were to be potent toxicant and fumigants against Sitophilus zeamais and Tribolium castaneum [116]. Rahman [117] reported that nicotine, an active component of Nicotiana tabacum, is a strong organic poison which acts as a contact-stomach poison with insecticidal properties. This compound is, of course, very toxic to humans as well. The essential oil vapours distilled from anise, cumin, eucalyptus, oregano, and rosemary were also reported as fumigantants and caused 100% mortality of the eggs of Tribolium confusum and Ephestia kuehniella [118]. Many species of the genus Ocimum oils, extracts, and their bioactive compounds have been reported to have insecticidal activities against various insect species [59, 119]. A list of many known toxicants from plant origin, reported as effective on stored-product insect-pest management, is given in Table 1.

Table 1: List of plant species reported to show insecticidal activity

Plant species	Family	**Plant part**	**References**
Acorus calamus	Acoraceae	O, RO	[27]
Allium sativum	Alliaceae	P	[28]
Annona squamosa	Annonaceae	L	[29]
Aphanamixis polystachya	Meliaceae	SC, SE	[25]
Azadirachta indica	Meliaceae	O, SP, LP	[30]
Baccharis salicifolia	Asteraceae	O	[31]
Bassia longifolia	Sapotaceae	E	[5]
Brassica spp.	Cruciferae	L, ZE	[32]
Cajanus cajan	Fabaceae	O	[33]
Calophyllum inophyllum	Clusiaceae	O	[34]
Calotropis procera	Apocynaceae	LP	[35]
Carum carvi	Apiaceae	FE	[36]
Cinnamomum aromaticum	Lauraceae	B	[37]
Citrus	Rutaceae	O	[38]

Curcuma longa	Zingiberaceae	P	[39]
Chenopodium ambrosioides	Amaranthaceae	FE, O	[40]
Cocos nucifera	Arecaceae	O	[25]
Convolvulus arvensis	Convolvulaceae	LE	[41]
Conyza dio scoridis	Asteraceae	ZE	[41]
Coriandrum sativum	Apiaceae	SE, O	[42]
Datura alba	Solanaceae	LP	[43]
Decalepis hamiltonii	Asclepiadaceae	XP	[7]
Eichhornia crassipes	Pontederiaceae	LE	[44]
Elae is guineensis	Arecaceae	O	[45]
Elae is guineensis	Palmaceae	O	[46]
Embelia ribes	Myrsinaceae	FE, O	[47]
Eucalyptus globules	Myrtaceae	LP, M	[48]
Foeniculum vulgare	Apiaceae	FE	[49]
Glycine max	Fabaceae	O	[50]
Jatropha gossypifolia	Euphorbiaceae	SE	[51]
Juniperus vi rginiana	Cupressaceae	O	[52]
Lantana camara	Verbenaceae	TE	[45]
Lonchocarpus spp.	Leguminosae	O	[53]
Lupinus albus	Fabaceae	SE	[54]
Lupinus termis	Leguminosae	SE	[54]
Melia azedarach	Meliaceae	O, E	[55]
Mentha citrate	Lamiaceae	O	[56]
Nicotiana tabacum	Solanaceae	E	[57]
Ocimum canum	Lamiaceae	LP	[58]
Ocimum kilimands charicum	Lamiaceae	O	[59]
Piper nigrum	Piperaceae	O, E	[60, 61]
Polygonum hydropiper	Polygonaceae	L	[62]
Pongamia glabra	Fabaceae	O, E	[61]
Psidium guajava	Myrtaceae	L, LP	[63]
Ryania speciosa	Flacourtiaceae	YE	[64]
Sapindus trifoliatus	Sapindaceae	SP	[65]
Schleichera trijuga	Sapindaceae	O	[66]
Sesamum orientale	Pedaliaceae	O	[5]
Sesamum indicum	Pedaliaceae	O	[67]

Syzy gium aromaticum	Myrtaceae	O	[68]
Tagetes erec ta	—	X, Y	[69]
Tanacetum cinerariaefolium	Asteraceae	O, P	[55]
Thujopsis dolabrata	Cupressaceae	E	[64]
Trigonella foenumgraecum	Fabaceae	SE	[70]
Vitex negundo	Lamiaceae	L	[71]

Note. L: leaves, B: bark, F: fruits, S: seeds, O: oil, P: powder, E: extract, M: vapour, R: Rhizome, T: plant, V: flower, X: root, and Y: stem, (Source: [5, 6]).

Natural Grain Protectants

From very early times, plant materials have been used as natural protectants of stored grains. Worldwide reports indicate that when mixed with stored grains, leaf, bark, seed powder, or oil extracts of plants reduce oviposition rate and suppress adult emergence of stored product insects, and also reduce seed damage rates [25, 40, 46, 87, 119–122]. In 1989, Jacobson [123] noted that the most promising natural grain protectants were generally observed in the plant families, Annonaceae, Asteraceae, Canellaceae, Labiatae, Meliaceae, and Rutaceae.

The Indian neem plant is the most well-known example and its various parts, namely, leaves, crushed seeds, powdered fruits, oil, and so forth, have been used to protect stored grain from infestation [1, 124, 125]. The neem oil and kernel powder gave effective grain protection against stored grain insect pests like Sitophilus oryzae, Tribolium cataneum, Rhyzopertha dominica, and Callosobruchus chinensis at the rate of 1 to 2% kernel powder or oil [126]. The neem oil adhered to grain forms uniform coating around the grains against storage pests for a period of 180–330 days [127]. Yadava and Bhatnagar [128] reported that a dried leaves ofAzadirachta indica have been mixed with stored grains for protection against insects. Azadirachtin is an active principle from the neem plant, which is an effective grain protectant against insect infestation [129]. Rajashekar et al. [7] reported that root powder extracts of Decalepis hamiltonii have been mixed with stored grains for protection against various stored grain insect pests. Eighteen species offered protection to wheat up to 9 months without affecting seed germination [98].

In parts of eastern Africa, leaves of some plants and allelochemicals including azadirachtin, nicotine, and rotenone have traditionally been used as grain protectants [5, 130]. The powders of Rauvolfia serpentina, Acorus calamus, and Mesua ferrea are used as a grain protectant against Rhyzopertha dominica [131]. In a survey in northern semiarid regions of Ghana only 16 plants were identified as being used as grain protectants [132]. In Africa, the grain protectant potential of different plant derivatives, including plant oils against major stored-product pests were also found to be very promising and reduced the risks associated with the use of insecticides [82, 121]. In northern Cameroon, the essential oils of plants Xylopia aethiopica, Vepris heterophylla, and Luppia rugosa are used for protection of stored grains from attack of stored grain insect pests [114]. The components of citrus peels were used as grain protectant against Callosobruchus maculatus [133]. Coconut oil has been found effective against Callosobruchus chinensis, for a storage period of six months, when applied to Vigna radiata (green gram) at 1% [134]. Formulations of menthol were used as protection of pulse grain from attack of Callosobruchus Chinensis [135]. Spinosad, a naturally occurring insecticide from the actinomycete, Saccharopolyspora spinosa, has high efficacy, a broad insect pest spectrum, low mammalian toxicity, and minimal environmental profile is unique among existing products currently used for stored-grain protection [136].

Chemosterilants/Reproduction Inhibitors

Many researchers reported that plant parts, oil, extracts, and powder mixed with grain-reduced insect oviposition, egg hatchability, postembryonic development, and progeny production [137–139]. Lists of 43 plant species have been reported as reproduction inhibitors against stored product insects [32]. Reports have also indicated that plant derivatives including the essential oils caused mortality of insect eggs [82]. Many ground plant parts, extracts, oils, and vapour also suppress many insects [6, 7].

Insect Growth and Development Inhibitors

Plant extracts showed deleterious effect on the growth and development of insects and reduced larval pupal and adult weight significantly, lengthened the larval and pupal periods, and reduced pupal recovery and adult eclosion [140]. Rajasekaran and Kumaraswami [141] reported that grains coated with plant extracts completely inhibited the development of insect like Sitophilus oryzae. Plant derivatives also reduce the survival rates of larvae and pupae and adult emergence [101]. Development of eggs and immature stages inside grain kernel were also inhibited by plant derivatives [102]. The crude extract also retarded development and caused mortality of larvae, cuticle melanisation, and high mortality in adults [142].

SOME IMPORTANT PHYTOCHEMICALS WITH INSECTICIDAL PROPERTIES

The botanical insecticides that have primarily been used and are commercially available include ryania, rotenone, pyrethrin, nicotine, azadirachtin, and sabadilla (Tables 2 and 3).

Table 2: List of insecticidal active principles of plants

Active principle	Plant species	Insect toxicity	Insect species	References
Anonaine	Annona reticulate	Contact	Callosobruchus chinensis	[72]
Azadirachtin	Azadirachta indica	Contact: IGR	Stored grain pests, aphids	[30]
E-Anethole	Foeniculum vulgare	Contact	Sitophilus oryzae, Callosobruchus chinensis	[49]
β-Asarone	Acorus calamus	Contact;	Stored grain pests	[73]
Z-Asarone	Acorus calamus; corus gramineus	Contact	Sitophilus zeamais	[26]

Bornyl acetate	Chamaecyparis obtuse	Contact	Sitophilus oryzae	[27]
Camphor	Ocimum kilimandscharicum	Contact	Sitophilus oryzae	[59]
(+)-3-Carene	Baccharis salicifolia	Contact	Tribolium castaneum	[59]
Carvacrol	Thujopsis dolabrata	Contact; fumigant	Sitophilus oryzae, Callosobruchus chinensis	[60]
Carvone	Carum carvi	Contact	Sitophilus oryzae, Rhyzopertha dominica	[74]
1,8 Cineole	Eucalyptus	Contact; fumigant	Rhyzopertha dominica Tribolium castaneum	[38]
Cinnamaldehyde	Cinnamomum aromaticum	Contact	Tribolium castaneum, Sitophilus zeamais	[37]
Dioctyl hexanedioate	Conyza dioscori dis	Contact	Tribolium castaneum, Sitophilus granaries	[41]
Eugenol	Citrus	Fumigant	Sitophilus oryzae	[38]
Estragole	Foeniculum vulgare	Contact	Sitophilus oryzae Lasioderma serricorne	[31]
(+)-Fenchone	Foeniculum vulgare	Contact	Sitophilus oryzae Lasioderma serricorne	[31]
Hexa decane	Chenopodiumambrosioides	Contact	Tribolium castaneum, Sitophilus granaries	[41]
Hexadecanoic acid	Convolvulus arvensis	Contact	Sitophilus oryzae, Rhyzopertha dominica.	[41]
Linalool	Ocimum canum Sims	Fumigant	Tribolium castaneum, Sitophilus granaries	[75]
Limonene	Citrus	Contact	Tribolium castaneum	[27]

(−)-Limonene	Baccharis salicifolia	Contact; fumigant	Tribolium castaneum	[31]
Nicotine	Nicotiana tabacum	Contact	Mites, aphids, thrips, leafhopper	[39]
Pyrethrin I and II	Tanacetum cinerariaefolium	Contact; stomach poison	Stored grain pests, crop pests	[76]
β-Pinene	Baccharis salicifolia	Contact	Tribolium castaneum,	[27]
α-Pinene	Baccharis salicifolia	Fumigant	Tribolium castaneum,	[27]
Rotenone	Lonchocarpus sp.	Contact; stomach poison	Crop pests, lace bugs, Sitophilus oryzae	[55]
Ryania	Ryania speciosa	Contact; stomach poison	Potato beetle, aphids, lace bugs, stored grain pests	[77]
Sabadilla	Schoenocaulon officinale	Contact; stomach poison	Stinks, thrips, squash bugs, leaf hoppers, caterpillars	[78]
Spinosyn A and D	Saccharopolyspora spinosa	Stomach poison	Stored grain pests	[79]

Table 3: Insecticidal activity and mammalian toxicity of some natural insecticides

Natural insecticides	**Insect toxicity***	**Mammalian toxicity Oral (rat) LD_{50} (mg/kg b.w.)**
Anethole	C, S	2090
β-asarone	C, S	275
Azadirachtin	IGR, R	13000
Carvacrol	C	810
1,8-Cineole	C, F	2480
Cinnamaldehyde	C	1160
Cuminic aldehyde	C, S	1390
Eugenol	C, F	500

Nicotine	C	50
Pyrethrin I and II	C, S	1200
Rotenone	S	350
Ryania	C, S	750
Sabadilla	C, S	5000
Spinosad	C	3738

*C: contact, S: stomach poison, F: fumigant, IGR: insect growth regulator, and R: repellent [80–82].

Ryania

The active components of ryania are derived from the roots and woody stems of the plant Ryania speciosa, native to Trinidad [143]. Ryania has low mammalian toxicity, with a median lethal dose (LD_{50}) of 750 mg/kg and works as both contact and stomach poison. It has long residual activity among the botanical insecticides. This botanical insecticide has a unique mode of action and affects muscles by binding to the calcium channels in the sarcoplasmic reticulum. This causes calcium ion flow into the cells, and death follows very rapidly [20]. Ryania works best on caterpillars (i.e., codling moth, corn earworm); however, is it also active on a wide range of insects and mites, including potato beetle, lace bugs, aphids and squash bug [144].

Rotenone

Rotenone is derived from the roots of the two plants: Lonchocarpus sp. and Derris sp. are both legumes originally from the East Indies, Malaya and South America. Rotenone is a moderately of the toxic botanical insecticides, with an LD_{50} of 132 mg/kg to mammals [81]. In fact, rotenone is more toxic to mammals than both carbaryl and malathion, two commonly used synthetically derived insecticides.

Also, rotenone is extremely toxic to fish [55]. This botanical insecticide works as both contact and stomach poison. Rotenone is slower acting than most other botanical insecticides, taking several days to kill pests; however, pests stop feeding almost immediately. It degrades rapidly in air and sunlight. Rotenone blocks respiration by electron transport on the complex I. Rotenone shows broad spectrum of activity on many insects and mite pests, including leaf-feeding beetles, caterpillars, lice, mosquitoes, ticks, fleas, and fire ants [145].

Pyrethrin/Pyrethrum

Pyrethrin I and II are derived from the seeds or flower of Chrysanthemum cinerariaefolium [55, 146] which is grown in Africa, Ecuador, and Kenya. Pyrethrin has a low mammalian toxicity. However, cats are highly susceptible to pyrethrin poisoning. The LD_{50} of pyrethrin is 1200 to 1,500 mg/kg [81, 147, 148]. Pyrethrin is one of the oldest household insecticides still available and is fast acting, providing almost immediate "knockdown" of insects following an application. It works as both a contact and a stomach poison. The material has a very short residual activity-degrading rapidly under sunlight, air and moisture, which means that frequent applications may be required. Pyrethrin can be used up until harvest, as there is no waiting interval required between initial application and harvest of food crops [149].

The way pyrethrin kills insects (mode of activity) is by disrupting the sodium and potassium ion-exchange process in insect nerves and interrupting the normal transmission of nerve impulses. Pyrethrin has activity on wide range of insects and mites, including flies, fleas, beetles, and spider mites [150].

Nicotine

Nicotine, which is derived from Nicotiana tabacum, is toxic to mammals among the botanical insecticides with an LD_{50} between 50 and 60 mg/kg [55, 151]. It is extremely harmful to humans. Nicotine, a fast-acting nerve toxin, works as a contact poison. It kills insects (and humans) through bonding to receptors at the nerve synapses (junctures), causing uncontrolled nerve firing, and by mimicking acetylcholine (Ach) at the nerve-muscle junctions in the central nervous system [152].

Certain plant types, such as roses, may be harmed or injured by nicotine sprays. Nicotine is most effective on soft-bodied insects and mites, including aphids, thrips, leafhoppers, and spider mites. Many caterpillars are resistant to nicotine [153].

Azadirachtin

Azadirachtin is derived from the tree Azadirachta indica, grown in India and Africa [55]. Azadirachtin has an extremely low mammalian toxicity and is least toxic of the commercial botanical insecticides, with an LD_{50} of 13,000 mg/kg. Azadirachtin is considered a contact poison; however, it has "some" systemic activity in plants when applied to the foliage. The material is generally nontoxic to beneficial insects and mites. Azadirachtin has broad mode of activity, working as a feeding deterrent, insect-growth regulator, repellent, and sterilant; and it may also inhibit oviposition [55, 154]. The material is active on a broad range of insects, including stored grain pests, aphids, caterpillars and mealybugs [30].

Sabadilla

Sabadilla is derived from the seeds of plant Schoenocaulon officinale, which is grown in Venezuela. Sabadilla is one of the least toxic registered botanical insecticides, with mammalian LD_{50} of 5,000 mg/kg. Sabadilla works as contact toxicant and a stomach poison. Similar to other botanical insecticides, the material has minimal residual activity and degrades rapidly in sunlight and moisture (rainfall). Sabadilla works by affecting nerve cell membranes, causing loss of nerve function, paralysis, and death [146]. It is effective against caterpillars, leaf hoppers, thrips, stink, and squash bugs.

Avermectins

Avermectins, which are macrocyclic lactones, are derived from the actinomycete, Streptomyces avermitilis[155], lethal dosage of 50% in range of 10–11.3 mg/kg for rat. This molecule is most effective against agricultural pests with lethal concentration of 90% (LC_{90}) in the range of 0.02 ppm for mites and has somewhat least toxicity to stored products pests. It is effective on internal parasites of domestic animals [156]. Avermectins block the neurotransmitter GABA at the neuromuscular junction in insects and mites. Visible activity, such as feeding and egg laying, stops shortly after exposure, though death may not occur for several days [157].

Spinosads

Spinosad is a mixture of spinosyn A and spinosyn D and was originally isolated from the soil Actinomycete,Saccharopolyspora spinosa [158]. Spinosad is recommended for the control of a very wide range of caterpillars, leaf miners and foliage-feeding beetles. Spinosyns have a novel mode of action, primarily targeting binding sites on nicotinic acetylcholine receptors that are distinct from those at which other insecticides exert activity, leading to disruption of acetylcholine neurotransmission [79, 159].

(Z) Asarone

(Z) Asarone is natural insecticide isolated from Acorus calamus L. [26]. This molecule is more effective against adults of Sitophilus oryzae, Lasioderma serricorne, and Callosobruchus chinensis and shows both fumigant and contact toxicity. Some studies show that the molecules possess in vivo carcinogenic effects [160] and in vitromutagenic activites [161]. Further, this molecule induces structural chromosome aberration in human lymphocytes in vitro [162]. Due to its mammalian toxicity [81, 163], the molecule is unsafe for grain treatment.

CHALLENGES TO THE UTILIZATION OF BOTANICALS PESTICIDES

Many plant species contain secondary metabolites that are potent against several pest species. Not only are some of the plants (e.g., the neem trees) of major interest as sources of phytochemicals for more environmentally benign grain/crop protection. Phytochemical products can increase income of rural farmers and promote safety and quality of food and life in general [8, 164].

The successful utilization of botanicals can help to control many of the world's destructive pests and diseases, as well as reduce erosion, desertification, deforestation, and perhaps even reducing human population by acting as spermaticide (although this will be considered a major negative effect by many cultures and religions) [165]. Although the possibilities of using botanical pesticides seem almost endless, many details remain to be clarified. Many obstacles must be overcome and many uncertainties clarified before their potential can be fully realized. These limitations seem surmountable; however, they present exciting challenges to the scientific and economic development communities. Solving the following obstacles and uncertainties may well bring a major new resource which will benefit much of the world. These obstacles include:

1. lack of experience and appreciation of the efficacy of botanicals for pest control. There are still doubts as to the effectiveness of plant-derived products (both "home-made" and commercial products) due to their slow action and lack of rapid knockdown effect;
2. genetic variability of plant species in different localities;
3. difficulty of registration and patenting of natural products and lack of standardization of botanical pesticide products;
4. economic uncertainties occasioned by seasonal supply of seeds, perennial nature of most botanical trees, and change in potency with location and time with respect to geographical limitations;
5. handling difficulties as there is no method for mechanizing the process of collecting, storing, or handling the seeds or leaves or flowers from some of the perennial trees;
6. instability of the active ingredients when exposed to direct sunlight;
7. competition with synthetic pesticides through aggressive advertising by commercial pesticides dealers and commercial-formulated botanicals are more expensive than synthetic insecticides and are not as widely available;
8. rapid degradation, although desirable in some respects, creates the need for more precise timing or more frequent applications;
9. Data on the effectiveness and long-term (chronic) mammalian toxicity are unavailable for some botanicals, and tolerances for some have not been established.

CONCLUSIONS

Many authors have evaluated the insecticidal (grain protectant) properties of plant products on various species of stored product insect pests. The results clearly show that it is possible to develop methods for grain protectants with reduced use of synthetic chemical insecticides. The main advantages of botanical pesticides are ecofriendly, easily biodegradable, nontoxic to nontarget organisms, and many plant-derived natural products acting against insects could be produced from locally available raw materials. They have been numerous botanical pesticides studied at the laboratory level. Research efforts should focus not only on their efficacy, but also on mammalian toxicity, mode of action in insects, seed germination, effect on nutritional quality, seedling growth, and stability of the compound. The insecticides of plant origin could be exploited for the development of novel molecules with highly precise targets for sustainable insect pest management in stored grain.

ACKNOWLEDGMENTS

The authors thank the Director, NBAII, Bangalore, for encouragement and support. The first author acknowledges the Department of Biotechnology, New Delhi, for awarding the postdoctoral research associateship.

REFERENCES

1. F. A. Talukder, M. S. Islam, M. S. Hossain, M. A. Rahman, and M. N. Alam, "Toxicity effects of botanicals and synthetic insecticides on Tribolium castaneum (Herbst) and Rhyzopertha dominica (F.),"Bangladesh Journal of Environment Science, vol. 10, no. 2, pp. 365–371, 2004.
2. N. K. Dubey, B. Srivastava, and A. Kumar. "Current status of plant products as botanical pesticides in storage pest management," Journal of Biopesticide, vol. 1, no. 2, pp. 182–186, 2008.
3. M. Jacobson, "Plants, insects, and man-their interrelationships," Economic Botany, vol. 36, no. 3, pp. 346–354, 1982

4. S. Ahmed and M. Grainge, "Potential of the neem tree (Azadirachta indica) for pest control and rural development," Economic Botany, vol. 40, no. 2, pp. 201–209, 1986
5. F. A. Talukder, "Plant products as potential stored product insect management agents–a mini review,"Emirates Journal of Agricultural Science, vol. 18, pp. 17–32, 2006.
6. S. Rajendran and V. Sriranjini, "Plant products as fumigants for stored-product insect control," Journal of Stored Products Research, vol. 44, no. 2, pp. 126–135, 2008
7. Y. Rajashekar, N. Gunasekaran, and T. Shivanandappa, "Insecticidal activity of the root extract ofDecalepis hamiltonii against stored-product insect pests and its application in grain protection," Journal of Food Science and Technology, vol. 47, no. 3, pp. 310–314, 2010
8. Y. Rajashekar and T. Shivanandappa, "A novel natural insecticide molecule for grain protection," inStored Products Protection, M. O. Carvalho, P. G. Fields, C. S. Adler, et al., Eds., Proceedings of the 10th International Working Conference on Stored Product Protection, pp. 913–917, Julius-Kühn-Archiv 425, 2010.
9. E. J. Donahaye, "Current status of non-residual control methods against stored product pests," Crop Protection, vol. 19, no. 8–10, pp. 571–576, 2000
10. B. C. Leelaja, Y. Rajashekar, P. Vanitha Reddy, K. Begum, and S. Rajendran, "Enhanced fumigant toxicity of allyl acetate to stored-product beetles in the presence of carbon dioxide," Journal of Stored Products Research, vol. 43, no. 1, pp. 45–48, 2007
11. Y. Rajashekar, P. V. Reddy, K. Begum, B. C. Leelaja, and S. Rajendran, "Studies on aluminium phosphide tablet formulation," Pestology, vol. 30, no. 4, pp. 41–45, 2006
12. WMO, "Scientific assessment of ozone depletion: 1994," Global Ozone Research and Monitoring Project37, Genoa, Italy, 1995.
13. G. P. Georghiou, "Over view of insecticide resistance," in Managing Resistance to Agrochemicals: From Fundamental Research to Practical Strategies, M. B. Green, H. M. Lebaron, and W. K. Moberg, Eds., vol. 421 of ACS Symposium Series, pp. 19–41, American Chemical Society, Washington, DC, USA, 1990.
14. B. R. Champ, "Occurrence of resistance to pesticides in grain storage pests," in Pesticides and Humid Tropical Grain Storage

System, B. R. Champ and E. Highly, Eds., vol. 14 of Proceedings of an International Seminar, Manila 1985. Canberra ACIAR proceedings, pp. 225–229, 1985.

15. B. Subramanyam and D. W. Hagstrum, "Resistance measurement and management," in Integrated Management of Insects in Stored Products, B. Subramanyam and D. W. Hagstrum, Eds., pp. 331–397, Marcel Dekker, New York, NY, USA, 1995.
16. S. C. Dubey, M. Suresh, and B. Singh, "Evaluation of Trichoderma species against Fusarium oxysporum f. sp. ciceris for integrated management of chickpea wilt," Biological Control, vol. 40, no. 1, pp. 118–127, 2007
17. R. Kumar, A. K. Mishra, N. K. Dubey, and Y. B. Tripathi, "Evaluation of Chenopodium ambrosioides oil as a potential source of antifungal, antiaflatoxigenic and antioxidant activity," International Journal of Food Microbiology, vol. 115, no. 2, pp. 159–164, 2007
18. K. Sharma and N. M. Meshram, "Bioactivity of essential oils from Acorus calamus and Syzygium aromaticum, against Sitophilus oryzae (L.) in stored wheat," Biopesticide International, vol. 2, pp. 144–152, 2006.
19. J. V. D. Heyde, R. C. Saxena, and H. Schmutterer, "Neem oil and neen extracts as potential insecticide for control of Hemipterous rice pests," in Proceedings of the 2nd International Neem Conference, pp. 337–390, Rauischholzhausen, Germany, 1984.
20. F. E. Dayan, C. L. Cantrell, and S. O. Duke, "Natural products in crop protection," Bioorganic and Medicinal Chemistry, vol. 17, no. 12, pp. 4022–4034, 2009
21. W. Hermawan, S. Nakajima, R. Tsukuda, K. Fujisaki, and F. Nakasuji, "Isolation of an antifeedant compound from Andrographis paniculata (Acanthaceae) against the diamond back, Plutella xylostella(Lepidoptera: Yponomeutidae)," Applied Entomology and Zoology, vol. 32, no. 4, pp. 551–559, 1997.
22. N. E. S. Lale, "A laboratory study of the comparative toxicity of products from three spices to the maize weevil," Postharvest Biology and Technology, vol. 2, no. 1, pp. 61–64, 1992
23. F. K. Ewete, J. T. Arnason, J. Larson, and B. J. R. Philogene, "Biological activities of extracts from traditionally used Nigerian

plants against the European corn borer, Ostrinia nubilalis," Entomologia Experimentalis et Applicata, vol. 80, no. 3, pp. 531–537, 1996

24. K. Sahayaraj, "Common plants oils in agriculture and storage pests management," Green Farming, vol. 1, no. 2, pp. 48–49, 2008.
25. F. A. Talukder and P. E. Howse, "Laboratory evaluation of toxic and repellent properiies of the pitharaj tree, Aphanamixis polstachya Wall & Parker, against Sitophilus oryzae (L.)," International Journal of Pest Management, vol. 40, pp. 274–279, 1995.
26. Y. Yingjuan, C. Wanlun, Y. Changju, X. Dong, and H. Yanzhang, "Isolation and characterization of insecticidal activity of (Z)-asarone from Acorus calamus (L.)," Insect Science, vol. 15, no. 3, pp. 229–236, 2008
27. C. Park, S. I. Kim, and Y. J. Ahn, "Insecticidal activity of asarones identified in Acorus gramineusrhizome against three coleopteran stored-product insects," Journal of Stored Products Research, vol. 39, no. 3, pp. 333–342, 2003
28. I. L. K. Park and S. C. Shin, "Fumigant activity of plant essential oils and components from garlic (Allium sativum) and clove bud (Eugenia caryophyllata) oils against the Japanese termite (Reticulitermes speratus Kolbe)," Journal of Agricultural and Food Chemistry, vol. 53, no. 11, pp. 4388–4392, 2005
29. J. A. Leatemia and M. B. Isman, "Toxicity and antifeedant activity of crude seed extracts of Annona squamosa (Annonaceae) against lepidopteran pests and natural enemies," International Journal of Tropical Insect Science, vol. 24, no. 2, pp. 150–158, 2004
30. E. D. Morgan, "Azadirachtin, a scientific gold mine," Bioorganic and Medicinal Chemistry, vol. 17, no. 12, pp. 4096–4105, 2009
31. M. García, O. J. Donadel, C. E. Ardanaz, C. E. Tonn, and M. E. Sosa, "Toxic and repellent effects ofBaccharis salicifolia essential oil on Tribolium castaneum," Pest Management Science, vol. 61, no. 6, pp. 612–618, 2005
32. F. A. Talukder, Isolation and characterization of the active secondary pithraj (Aphanamixis polystachya) compounds in controlling stored-product insect pests [Ph.D. thesis], University of Southampton, Southampton, UK, 1995.

33. E. R. Prasad, H. Merzendorfer, C. Madhurarekha, A. Dutta-Gupta, and K. Padmasree, "Bowman-birk proteinase inhibitor from Cajanus cajan seeds: purification, characterization, and insecticidal properties," Journal of Agricultural and Food Chemistry, vol. 58, no. 5, pp. 2838–2847, 2010

34. A. F. Traboulsi, K. Taoubi, S. El-Haj, J. M. Bessiere, and S. Rammal, "Insecticidal properties of essential plant oils against the mosquito Culex pipiens molestus (Diptera: Culicidae)," Pest Management Science, vol. 58, no. 5, pp. 491–495, 2002

35. M. Singh, K. Lal, and S. B. Singh, "Effect of calotropis (Calotropis procera) extract on infestation of termite (Odontotermes obesus) in sugarcane hybrid," Indian Journal of Agricultural Sciences, vol. 72, no. 7, pp. 439–441, 2002

36. M. C. Digilio, E. Mancini, E. Voto, and V. De Feo, "Insecticide activity of Mediterranean essential oils,"Journal of Plant Interactions, vol. 3, no. 1, pp. 17–23, 2008

37. I. K. Park, H. S. Lee, S. G. Lee, J. D. Park, and Y. J. Ahn, "Insecticidal and fumigant activities ofCinnamomum cassia bark-derived materials against Mechoris ursulus (Coleoptera: Attelabidae),"Journal of Agricultural and Food Chemistry, vol. 48, no. 6, pp. 2528–2531, 2000

38. H. T. Prates, J. P. Santos, J. M. Waquil, J. D. Fabris, A. B. Oliveira, and J. E. Foster, "Insecticidal activity of monoterpenes against Rhyzopertha dominica (F.) and Tribolium castaneum (Herbst)," Journal of Stored Products Research, vol. 34, no. 4, pp. 243–249, 1998

39. A. K. Tripathi, V. Prajapati, N. Verma et al., "Bioactivities of the leaf essential oil of Curcuma longa (var. ch-66) on three species of stored-product beetles (Coleoptera)," Journal of Economic Entomology, vol. 95, no. 1, pp. 183–189, 2002

40. L. A. Tapondjou, C. Adler, H. Bouda, and D. A. Fontem, "Efficacy of powder and essential oil fromChenopodium ambrosioides leaves as post-harvest grain protectants against six-stored product beetles,"Journal of Stored Products Research, vol. 38, no. 4, pp. 395–402, 2002

41. G. S. Peterson, M. A Kandil, M. D. Abdallah, and A. A. A. Fraq, "Isolation and characterization of biologically-active compounds

from some plant extracts," Pesticide Science, vol. 25, pp. 337–342, 1989.

42. M. S. Islam, M. M. Hasan, W. Xiong, S. C. Zhang, and C. L. Lei, "Fumigant and repellent activities of essential oil from Coriandrum sativum (L.) (Apiaceae) against red flour beetle Tribolium castaneum(Herbst) (Coleoptera: Tenebrionidae)," Journal of Pest Science, vol. 82, no. 2, pp. 171–179, 2009

43. S. Ahamed, F. Shakil, M. A. Riaz, and A. Hussain, "Comparative efficacy of Datura alba nees, Calotropis procera and Imidacloprid on termites in sugarcane at Faisalabad," Pakistan Entomology, vol. 27, no. 2, pp. 11–14, 2005.

44. R. Charudattan, "Integrated control of water hyacinth Eichhornia crassipes with a pathogen, insects, and herbicides," Weed Science, vol. 34, pp. 26–30, 1986.

45. H. Bouda, L. A. Tapondjou, D. A. Fontem, and M. Y. D. Gumedzoe, "Effect of essential oils from leaves of Ageratum conyzoides, Lantana camara and Chromolaena odorata on the mortality of Sitophilus zeamais (Coleoptera, Curculionidae)," Journal of Stored Products Research, vol. 37, no. 2, pp. 103–109, 2001

46. D. Obeng-Ofori, "Plant oils as grain protectants against infestations of Cryptolestes pusillus andRhyzopertha dominica in stored grain," Entomologia Experimentalis et Applicata, vol. 77, no. 2, pp. 133–139, 1995

47. S. Ahmed and M. Grainge, "Potential of the neem tree (Azadirachta indica) for pest control and rural development," Economic Botany, vol. 40, no. 2, pp. 201–209, 1986

48. D. R. Batish, H. P. Singh, R. K. Kohli, and S. Kaur, "Eucalyptus essential oil as a natural pesticide," Forest Ecology and Management, vol. 256, no. 12, pp. 2166–2174, 2008

49. D. H. Kim and Y. J. Ahn, "Contact and fumigant activities of constituents of Foeniculum vulgare fruit against three coleopteran stored-product insects," Pest Management Science, vol. 57, no. 3, pp. 301–306, 2001

50. J. H. Pankey, J. L. Griffin, B. R. Leonard, D. K. Miller, R. G. Downer, and R. W. Costello, "Glyphosate-insecticide combination effects on weed and insect control in cotton," Weed Technology, vol. 18, no. 3, pp. 698–703, 2004

51. S. Phowichit, S. Buatippawan, and V. Bullangpoti, "Insecticidal activity of Jatropha gossypifolia L. (Euphorbiaceae) and Cleome viscosa L. (Capparidacae) on Spodoptera litura (Lepidoptera: Noctuidae). Toxicity and carboxylesterase and glutathione-S-transferase activities studies," Communications in Agricultural and Applied Biological Sciences, vol. 73, no. 3, pp. 611–619, 2008.
52. J. R. Sabine, "Exposure to an environment containing the aromatic red cedar, Juniperus virginiana: procarcinogenic, enzyme inducing and insecticidal effects," Toxicology, vol. 5, no. 2, pp. 221–235, 1975.
53. M. Isman, "Pesticides based on plant essential oils," Pesticide Outlook, vol. 10, no. 2, pp. 68–72, 1999
54. K. Bermúdez-Torres, J. Martínez Herrera, R. Figueroa Brito, M. Wink, and L. Legal, "Activity of quinolizidine alkaloids from three Mexican Lupinus against the lepidopteran crop pest Spodoptera frugiperda," BioControl, vol. 54, no. 3, pp. 459–466, 2009
55. M. B. Isman, "Botanical insecticides, deterrents, and repellents in modern agriculture and an increasingly regulated world," Annual Review of Entomology, vol. 51, pp. 45–66, 2006
56. R. Pavela, "Insecticidal activity of some essential oils against larvae of Spodoptera littoralis," Fitoterapia, vol. 76, no. 7-8, pp. 691–696, 2005
57. A. Tiwari, M. Lakshamana Kumar, and R. C. Saxena, "Effect of Nicotiana tabacum on Tribolium castaneum," International Journal of Pharmacognosy, vol. 33, no. 4, pp. 348–350, 1995
58. A. Delobel and P. Malonga, "Insecticidal properties of six plant materials against Caryedon serratus (Ol.) (Coleoptera: Bruchidae)," Journal of Stored Products Research, vol. 23, no. 3, pp. 173–176, 1987
59. D. Obeng-Ofori, C. H. Reichmuth, A. J. Bekele, and A. Hassanali, "Toxicity and protectant potential of camphor, a major component of essential oil of Ocimum kilimandscharicum, against four stored product beetles," International Journal of Pest Management, vol. 44, no. 4, pp. 203–209, 1998
60. H. C. F. Su and R. Horvat, "Isolation, identification, and insecticidal properties of Piper nigrum amides,"Journal of Agricultural and Food Chemistry, vol. 29, no. 1, pp. 115–118, 1981

61. S. Sighamony, I. Anees, T. Chandrakala, and Z. Osmani, "Efficacy of certain indigenous plant products as grain protectants against Sitophilus oryzae (L.) and Rhyzopertha dominica (F.)," Journal of Stored Products Research, vol. 22, no. 1, pp. 21–23, 1986

62. I. Rahman, I. Gogoi, and S. P. Biswas, "Insecticidal activity of leaf extracts of Polygonum hydropiperLinn. against Odontotermes assamensis Holm. (Isoptera:Termitidae)," Journal of Entomological Research, vol. 6, no. 4, pp. 379–383, 2007.

63. A. Sharaby, "Evaluation of some Myrtaceae plant leaves as protectants against the infestation bySitophilus oryzae L. and Sitophilus granaries (L.)," Insect Science and its Application, vol. 9, no. 4, pp. 465–468, 1988.

64. L. J. Charles, "Toxicity of the botanical insecticide Ryania speciosa to Culex pipiens fatigans Wied,"Bulletin of Entomological Reseearch, vol. 45, pp. 403–410, 1954.

65. S. S. Rahman, M. M. Rahman, S. A. Begum, M. M. R. Khan, and M. M. H. Bhuiyan, "Investigation ofSapindus mukorossi extracts for repellency, insecticidal activity and plant growth regulatory effect,"Journal of Applied Science Research, vol. 3, no. 2, pp. 95–101, 2007.

66. K. L. Mikolajczak, R. V. Madrigal, C. R. Smith, and D. K. Reed, "Insecticidal effects of cyanolipids on three species of stored product insects, European corn borer (Lepidoptera: Pyralidae) larvae, and striped cucumber beetle (Coleoptera: Chrysomelidae)," Journal of Economic Entomology, vol. 77, no. 5, pp. 1144–1148, 1984.

67. H. F. Khater and D. F. Khater, "The insecticidal activity of four medicinal plants against the blowflyLucilia sericata (Diptera: Calliphoridae)," International Journal of Dermatology, vol. 48, no. 5, pp. 492–497, 2009

68. M. B. Isman, "Botanical insecticides: for richer, for poorer," Pest Management Science, vol. 64, no. 1, pp. 8–11, 2008

69. F. Nikkon, M. R. Habib, M. R. Karim, Z. Ferdousi, M. M. Rahman, and M. E. Haque, "Insecticidal activity of flower ofTagetes erecta L. against Tribolium castaneum (Herbst)," Research Journal of Agricultural and Biological Sciences, vol. 5, no. 5, pp. 748–753, 2009.

70. J. Pemonge, M. J. Pascual-Villalobos, and C. Regnault-Roger, "Effects of material and extracts ofTrigonella foenum-graecum L. against the stored product pests Tribolium castaneum (Herbst) (Coleoptera: Tenebrionidae) and Acanthoscelides obtectus (say) (Coleoptera: Bruchidae)," Journal of Stored Products Research, vol. 33, no. 3, pp. 209–217, 1997.

71. Y. Lin, X. Ming, X. Jian, and L. Chang-Hoo, "Toxicity of Vitex negando extracts to several insect pests,"Pesticides, pp. 2–6, 2004.

72. B. Oliver-Bever, Medicinal Plants in Tropical West Africa, Cambridge University Press, Cambridge, UK, 1986.

73. R. M. Baxter, P. C. Dandiya, S. I. Kandel, A. Okany, and G. C. Walker, "Separation of the hypnotic potentiating principles from the essential oil of Acorus Calamus L. of Indian origin by liquid-gas chromatography," Nature, vol. 185, no. 4711, pp. 466–467, 1960

74. F. A. Afifi, M. Salem, and A. M. Hekal, "Insecticidal properties of the extracts of lupin seed and caraway fruits against some stored product insects," Annals of Agricultural Science, vol. 34, no. 1, pp. 401–414, 1989.

75. D. K. Weaver, F. V. Dunkel, L. Ntezurubanza, L. L. Jackson, and D. T. Stock, "The efficacy of linalool, a major component of freshly-milled Ocimum canum Sims (Lamiaceae), for protection against postharvest damage by certain stored product Coleopteran," Journal of Stored Products Research, vol. 27, no. 4, pp. 213–220, 1991

76. F. Tattersfield, R. P. Hobson, and C. T. Gimingham, "Pyrethrin I and II: their insecticidal valve and estimation in Pyrethrum (Chrysanthemum cinebariaefolium). I," Journal of Agricultural Science, vol. 19, pp. 266–296, 1929.

77. P. R. Jefferies, R. F. Toia, B. Brannigan, I. Pessah, and J. E. Casida, "Ryania insecticide: analysis and biological activity of 10 natural ryanoids," Journal of Agricultural and Food Chemistry, vol. 40, no. 1, pp. 142–146, 1992

78. J. D. Hare and J. G. Morse, "Toxicity, persistence and potency of sabadilla alkaloid formulations to Citrus thrips (Thysanoptera: Thripidae)," Journal of Economic Entomology, vol. 90, no. 2, pp. 326–332, 1997

79. T. C. Sparks, G. D. Crouse, and G. Durst, "Natural products as insecticides: the biology, biochemistry and quantitative structure-activity relationships of spinosyns and spinosoids," Pest Management Science, vol. 57, no. 10, pp. 896–905, 2001

80. F. A. Khater, "Prospects of botanical biopesticides in insect pest management," Journal of Applied Pharmaceutical Science, vol. 2, no. 9, pp. 244–259, 2012.

81. P. Golob and I. Gudrups, "The use of spices and medicinals as bioactive protectants for grains," FAO Agricultural Sciences Bulletin 137, FAO, Rome, Italy, 1999.

82. D. Obeng-Ofori and C. Reichmuth, "Bioactivity of eugenol, a major component of essential oil ofOcimum suave (Wild.) against four species of stored-product Coleopteran," International Journal of Pest Management, vol. 43, no. 1, pp. 89–94, 1997

83. A. A. Abdel-Gawad and H .A. Khatab, "Soil and plant protection methods in ancient Egypt," inProceedings of the 2nd International Conference on Soil Pollution, pp. 19–22, 1985.

84. J. Varma and N. K. Dubey, "Prospectives of botanical and microbial products as pesticides of tomorrow," Current Science, vol. 76, no. 2, pp. 172–179, 1999

85. A. Hassanalli and W. Lwande, "Antipest secondary metabolites from African plants," in Insecticides of Plant Origin, J. T. Arnason, B. J. R. Philogene, and P. Morand, Eds., ACS Symposium Series, pp. 78–94, American Chemical Society, Washington, DC, USA, 1989.

86. Y. Akhtar, Y. R. Yeoung, and M. B. Isman, "Comparative bioactivity of selected extracts from Meliaceae and some commercial botanical insecticides against two noctuid caterpillars, Trichoplusia ni andPseudaletia unipuncta," Phytochemistry Reviews, vol. 7, no. 1, pp. 77–88, 2008

87. A. K. Tripathi, S. Upadhyay, M. Bhuiyan, and P. R. Bhattacharya, "A review on prospects of essential oils as biopesticides in insect-pest management," Journal of Pharmocological Phytotheraphy, vol. 1, no. 5, pp. 52–63, 2009.

88. S. Ahmed and B. Koppel, "Plant extracts for pest control: village level processing and use by limited resource farmers," in Proceedings of the American Association for the Advancement of Science Annual Meeting, Los Angeles, Calif, USA, May 1985.

89. J. L. Wolfson, R. E. Shade, P. E. Mentzer, and L. L. Murdock, "Efficacy of ash for controlling infestations of Callosobruchus maculatus (F.) (Coleoptera: Bruchidae) in stored cowpeas," Journal of Stored Products Research, vol. 27, no. 4, pp. 239–243, 1991.
90. R. G. Powel, "Higher plants as a source of new insecticidal compounds," Pesticide Science, vol. 27, pp. 228–229, 1989.
91. E. O. Owusu, "Effect of some Ghanaian plant components on control of two stored-product insect pests of cereals," Journal of Stored Products Research, vol. 37, no. 1, pp. 85–91, 2001
92. R. C. Saxena, G. Jillani, and A. A. Kareem, "Effects of neem on stored grain insects," in Focus on Phytochemical Pesticides, Volume 1: The Neem Tree, M. Jacobson, Ed., pp. 97–111, CRC Press, Boca Raton, Fla, USA, 1988.
93. A. Chatterjee, B. Das, N. Adityachaudhury, and S. Debkirtaniya, "Note on insecticidal properties of the seeds of Jatropha gossypifolia Linn," Indian Journal of Agricultural Science, vol. 50, no. 8, pp. 637–638, 1980.
94. M. J. Sim, D. R. Choi, and Y. J. Ahn, "Vapor phase toxicity of plant essential oils to Cadra cautella(Lepidoptera: Pyralidae)," Journal of Economic Entomology, vol. 99, no. 2, pp. 593–598, 2006.
95. G. Singh and K. Upadhyay, "Essential oils: a potent source of natural pesticides," Journal of Scientific and Industrial Research, vol. 52, pp. 676–683, 1993.
96. M. Mariappan, S. Jayaraj, and R. C. Saxena, "Effect of nonedible seed oils on survival of Nephotettix virescens (Homoptera: Cicadellidae) and on transmission of rice tungro virus," Journal of Economic Entomology, vol. 81, no. 5, pp. 1369–1372, 1988.
97. R. C. Saxena, O. P. Dixit, and P. Sukumaran, "Laboratory assessment of indigenous plant extracts for anti-juvenile hormone activity in Culex quinquefasciatus," Indian Journal of Medical Research A, vol. 95, pp. 204–206, 1992
98. K. C. Devi and S. S. Devi, "Insecticidal and oviposition deterrent properties of some spices against coleopteran beetle, Sitophilus oryzae," Journal of Food Science and Technology, pp. 1–5, 2011
99. M. F. Maia and S. J. Moore, "Plant-based insect repellents: a review of their efficacy, development and testing," Malaria Journal, vol. 10, no. 1, article S11, pp. 1–15, 2011.

100. Y. S. Xie, P. G. Fields, and M. B. Isman, "Repellency and toxicity of azadirachtin and neem concentrates to three stored-product beetles," Journal of Economic Entomology, vol. 88, no. 4, pp. 1024–1031, 1995.
101. O. Koul, S. Walia, and G. S. Dhaliwal, "Essential oils as green pesticides: potential and constraints,"Biopesticide International, vol. 4, no. 1, pp. 63–88, 2008.
102. S. J. Boeke, I. R. Baumgart, J. J. A. Van Loon, A. Van Huis, M. Dicke, and D. K. Kossou, "Toxicity and repellence of African plants traditionally used for the protection of stored cowpea againstCallosobruchus maculatus," Journal of Stored Products Research, vol. 40, no. 4, pp. 423–438, 2004.
103. A. K. Tripathi, V. Prajapati, A. Ahmad, K. K. Aggarwal, and S. P. S. Khanuja, "Piperitenone oxide as toxic, repellent, and reproduction retardant toward malarial vector Anopheles stephensi (Diptera: Anophelinae)," Journal of Medical Entomology, vol. 41, no. 4, pp. 691–698, 2004.
104. K. Munakata, "Insect antifeedants of Spodoptera litura in plants," in Host Plant Resistance to Pests, P. A. Hedin, Ed., vol. 62 of ACS Symposium Series, pp. 185–196, American Chemical Society, Washington, DC, USA, 1997.
105. R. C. Saxena, G. Jilani, and A. A. Kareem, "Effects of neem on stored grain insects. Focus on phytochemical pesticides," Florida Entomology, vol. 1, pp. 97–111, 1988.
106. F. A. Talukder and P. E. Howse, "Isolation of secondary plant compounds from Aphanamixis polystachya as feeding deterrents against adult Tribolium castaneum (Coleoptera: Tenebrionidae),"Journal of Stored Products Research, vol. 107, no. 5, pp. 395–402, 2000.
107. L. A. Hummelbrunner and M. B. Isman, "Acute, sublethal, antifeedant, and synergistic effects of monoterpenoid essential oil compounds on the tobacco cutworm, Spodoptera litura (Lep., Noctuidae),"Journal of Agricultural and Food Chemistry, vol. 49, no. 2, pp. 715–720, 2001.
108. Z. L. Liu, Y. J. Xu, J. Wu, S. H. Goh, and S. H. Ho, "Feeding deterrents from Dictamnus dasycarpusTurcz against two stored-product insects," Journal of Agricultural and Food Chemistry, vol. 50, no. 6, pp. 1447–1450, 2002.

109. H. C. F. Su and R. Horvat, "Isolation and characterization of four major components from insecticidally active lemon peel extract," Journal of Agricultural and Food Chemistry, vol. 35, no. 4, pp. 509–511, 1987.

110. H. C. F. Su, "Toxicity and repellency of Chenopodium oil to four species of stored-product insects," Journal of Entomological Science, vol. 26, pp. 178–182, 1991.

111. D. K. Weaver, F. V. Dunkel, L. Ntezurubanza, L. L. Jackson, and D. T. Stock, "The efficacy of linalool, a major component of freshly-milled Ocimum canum Sims (Lamiaceae), for protection against postharvest damage by certain stored product Coleopteran," Journal of Stored Products Research, vol. 27, no. 4, pp. 213–220, 1991.

112. A. K. Tripathi, V. Prajapati, K. K. Aggarwal, S. P. S. Khanuja, and S. Kumar, "Repellency and toxicity of oil from Artemisia annua to certain stored-product beetles," Journal of Economic Entomology, vol. 93, no. 1, pp. 43–47, 2000.

113. C. Channoo, S. Tantakom, S. Jiwajinda, and S. Isichaikul, "Fumigation toxicity of eucalyptus oil against three stored-product beetles," Thailand Journal of Agricultural Science, vol. 35, no. 5, pp. 265–272, 2002.

114. T. S. L. Ngamo, I. Ngatanko, M. B. Ngassoum, P. M. Mapongmestsem, and T. Hance, "Persistence of insecticidal activities of crude essential oils of three aromatic plants towards four major stored product insect pests," African Journal of Agricultural Research, vol. 2, pp. 173–177, 2007.

115. M. J. Pascual-Villalobos and A. Robledo, "Screening for anti-insect activity in Mediterranean plants,"Industrial Crops and Products, vol. 8, no. 3, pp. 183–194, 1998.

116. Y. Huang, S. X. Chen, and S. H. Ho, "Bioactivities of methyl allyl disulfide and diallyl trisulfide from essential oil of garlic to two species of stored-product pests, Sitophilus zeamais (Coleoptera: Curculionidae) and Tribolium castaneum (Coleoptera: Tenebrionidae)," Journal of Economic Entomology, vol. 93, no. 2, pp. 537–543, 2000.

117. M. D. M. Rahman, "Some promising physical, botanical and chemical methods for the protection of grain legumes against bruchids in storage under Bangladesh conditions," in Bruchids

and Legumes: Economics, Ecology and Coevolution, K. Fujii, A. M. R. Gatehouse, C. D. Johnson, R. Mitchell, and T. Yoshida, Eds., pp. 63–73, Kluwer Academic Publishers, Dordrecht, The Netherlands, 1990.

118. I. Tunç, B. M. Berger, F. Erler, and F. Dagli, "Ovicidal activity of essential oils from five plants against two stored-product insects," Journal of Stored Products Research, vol. 36, no. 2, pp. 161–168, 2000.

119. S. M. Kéita, C. Vincent, J. P. Schmit, J. T. Arnason, and A. Bélanger, "Efficacy of essential oil of Ocimum basilicum L. and O. gratissimum L. applied as an insecticidal fumigant and powder to controlCallosobruchus maculatus (Fab.) (Coleoptera: Bruchidae)," Journal of Stored Products Research, vol. 37, no. 4, pp. 339–349, 2001.

120. I. Onu and M. Aliyu, "Evaluation of powdered fruits of four peppers (Capsicum spp.) for the control ofCallosobruchus maculatus (F) on stored cowpea seed," International Journal of Pest Management, vol. 41, no. 3, pp. 143–145, 1995.

121. E. Shaaya, M. Kostjukovski, J. Eilberg, and C. Sukprakarn, "Plant oils as fumigants and contact insecticides for the control of stored-product insects," Journal of Stored Products Research, vol. 33, no. 1, pp. 7–15, 1997.

122. F. Bakkali, S. Averbeck, D. Averbeck, and M. Idaomar, "Biological effects of essential oils–a review,"Food and Chemical Toxicology, vol. 46, no. 2, pp. 446–475, 2008.

123. M. Jacobson, "Botanical pesticides: past, present and future," in Insecticides of Plant Origin, J. J. Arnason, B. R. Philogen, and P. Morand, Eds., vol. 387 of ACS Symposium Series, pp. 1–10, Washington, DC, USA, 1989.

124. M. G. Jotwani P. Sircar, "Neem seed as protectant against stored grain pests infesting wheat seed,"Indian Journal of Entomology, vol. 27, pp. 160–164, 1965.

125. D. A. Devi and N. Mohandas, "Relative efficacy of some antifeedants and deterrents against insect pests of stored paddy," Entomon, vol. 7, no. 3, pp. 261–264, 1982.

126. J. Pereira and R. Wohlgemuth, "Neem (Azadirachta indica A. Juss) of West African origin as a protectant of stored maize," Journal of Applied Entomology, vol. 94, pp. 208–214, 1982.

127. S. M. Ahmed, "Neem (Azadirachta Indica A. Juss) a safer insecticide potentials and prospects," Pest Management, pp. 1–3, 1994.
128. S. R. S. Yadava and K. N. Bhatnagar, "A preliminary study on the protection of stored cowpea grains against pulse beetle, by indigenous plant products," Pesticides, vol. 21, no. 8, pp. 25–29, 1987.
129. H. Schmutterer, "Properties and potential of natural pesticides from the neem tree, Azadirachta indica,"Annual Review of Entomology, vol. 35, pp. 271–297, 1990.
130. A. Hassanali, W. Lwande, N. Ole-Silayo, L. Moreka, S. Nokoe, and A. Chapya, "Weevil repellent constituents of Ocimum suave leaves and Eugenia caryophyllata cloves used as grain protectants in parts of Eastern Africa," Discovery and Innovation, vol. 2, no. 2, pp. 91–95, 1990.
131. S. N. Tiwari, "Efficacy of some plant products as grain protectants against Rhizopertha dominica (F.) (Coleoptera; Bostrichidae)," International Journal of Pest Management, vol. 40, no. 1, pp. 94–97, 1994.
132. J. Brice, C. Moss, N. Marsland et al., "Post-harvest constraints and opportunities in cereal and legume production production system in northern Ghana," NRI Research Report 85, 1996.
133. K. N. Don-Pedro, "Toxicity of some citrus peels to Dermestes maculatus Deg. and Callosobruchus maculatus (F)," Journal of Stored Products Research, vol. 21, no. 1. pp. 31–34, 1985
134. R. B. Doharey, R. N. Katiyar, and K. M. Singh, "Eco-toxicological studies on pulse beetles infesting green gram," Bulletin of Grain Technology, vol. 28, no. 2, pp. 116–119. 1990.
135. D. Singh and S. S. Mehta, "Menthol containing formulation inhibits adzuki bean beetle, Callosobruchus chinensis l. (Coleoptera; Bruchidae) population in pulse grain storage," Journal of Biopesticides, vol. 3, no. 3, pp. 596–603, 2010
136. M. B. Hertlein, G. D. Thompson, B. Subramanyam, and C. G. Athanassiou, "Spinosad: a new natural product for stored grain protection," Journal of Stored Products Research, vol. 47, no. 3, pp. 131–146, 2011

137. B. P. Saxena, K. Tikku, C. K. Atal, and O. Koul, "Insect antifertility and antifeedant allelochemics inAdhatoda vasica," Insect Science and its Application, vol. 7, no. 4, pp. 489–493, 1986.

138. E. F. Asawalam and S. O. Adesiyan, "Potentials of Ocimum basilicum (Linn.) for the control of Sitophilus zeamais (Motsch)," The Nigerian Agricultural Journal, vol. 32, pp. 195–201, 2001.

139. G. H. Schmidt, N. M. M. Ibrahim, and M. D. Abdallah, "Toxicological studies on the long-term effects of heavy metals (Hg, Cd, Pb) in soil on the development of Aiolopus thalassinus (Fabr.) (Saltatoria: Acrididae)," Science of the Total Environment, vol. 107, pp. 109–133, 1991

140. L. A. M. Khanam, D. Talukder, A. R. Khan, and S. M. Rahman, "Insectidical properties of Royna,Aphanamixis polystachya Wall. (Parker) (Meliaceae) against Tribolium confusum Duval," Journal of Asiatic Society of Bangladesh Science, vol. 16, pp. 71–74, 1990.

141. B. Rajasekaran and T. Kumaraswami, "Studies on increasing the efficacy on neem seed kernel extract," in Behavioural and Physiological Approaches in Pest Management, A. Regupathy and S. Jayaraj, Eds., pp. 29–30, Khadi and Village Industries Commission, Pune, India, 1985.

142. K. Jamil, U. Rani, and G. Thyagarajan, "Water hyacinth–a potential new juvenile hormone mimic,"International Pest Control, vol. 26, no. 4, pp. 106–108, 1984.

143. B. P. Pepper and L. A. Carruth, "A new plant insecticide for control of the European corn borer,"Journal of Economic Entomology, vol. 38, pp. 59–66, 1945.

144. N. Z. Dimetry, "Prospects of botanical pesticides for the future in integrated pest management programme (IPM) with special reference to neem uses in Egypt," Archives of Phytopathology and Plant Protection, vol. 45, no. 10, pp. 1138–1161, 2012.

145. W. M. Davidson, "Rotenone as a contact insecticide," Journal of Economic Entomology, vol. 23, no. 5, pp. 868–874, 1930.

146. L. G. Copping and S. O. Duke, "Natural products that have been used commercially as crop protection agents," Pest Management Science, vol. 63, no. 6, pp. 524–554, 2007

147. J. Miyamoto, "Degradation, metabolism and toxicity of synthetic pyrethroids," Environmental Health Perspectives, vol. 14, pp. 15–28, 1976

148. J. E. Brooks, P. J. Savarie, and J. J. Johnston, "The oral and dermal toxicity of selected chemicals to brown tree snakes (Boiga irregularis)," Wildlife Research, vol. 25, no. 4, pp. 427–435, 1998.

149. J. E. Casida and G. B Quistad, Pyrethrum Flowers: Production, Chemistry, Toxicology and Uses, Oxford University Press, Oxford, UK, 1995.

150. J. E. Casida, "Pyrethrum flowers and pyrethroid insecticides," Environmental Health Perspectives, vol. 34, pp. 189–202, 1980.

151. I. Yamomoto, "Nicotine to nicotinoids," in Nicotinoid Insecticides and the Nicotinic Acetylcholine Receptor, I. Yamomoto and J. E Casida, Eds., Springer, Tokoyo, Japan, 1999.

152. I. Ujvary, "Nicotine and other alkaloids," in Nicotinoid Insecticides and the Nicotinic Acetylcholine Receptor, I. Yamomoto and J. E Casida, Eds., pp. 29–70, Springer, Tokoyo, Japan, 1999.

153. N. Zong and C. Wang, "Induction of nicotine in tobacco by herbivory and its relation to glucose oxidase activity in the labial gland of three noctuid caterpillars," Chinese Science Bulletin, vol. 49, no. 15, pp. 1596–1601, 2004.

154. H. Rembold, "Azadirachtins: their structure and mode of action," in Insecticides of Plant Origin, J. T. Arnason, B. J. R. Philogene, and P. Morand, Eds., ACS Symposium series, pp. 150–163, American Chemical Society, Washington DC, USA, 1989.

155. I. Putter, J. G. Mac Connell, and F. A. Preiser, "Avermectins: novel insecticides, acaricides and nematicides from a soil microorganism," Cellular and Molecular Life Sciences, vol. 37, no. 9, pp. 963–964, 1981.

156. H. Mrozik, P. Eskola, B. O. Linn et al., "Discovery of novel avermectins with unprecedented insecticidal activity," Cellular and Molecular Life Sciences, vol. 45, no. 3, pp. 315–316, 1989.

157. Y. Deng, "House fly brain γ-aminobutyric acid-gated chloride channel: target for multiple classes of insecticides," Pesticide Biochemistry and Physiology, vol. 41, no. 1, pp. 60–65, 1991.

158. L. G. Copping and J. J. Menn, "Biopesticides: a review of their action, applications and efficacy," Pest Management Science, vol. 56, pp. 651–676, 2000.

159. D. E. Snyder, J. Meyer, A. G. Zimmermann et al., "Preliminary studies on the effectiveness of the novel pulicide, spinosad, for the treatment and control of fleas on dogs," Veterinary Parasitology, vol. 150, no. 4, pp. 345–351, 2007.

160. R. W. Wiseman, E. C. Miller, J. A. Miller, and A. Liem, "Structure-activity studies of the hepatocarcinogenicities of alkenylbenzene derivatives related to estragole and safrole on administration to preweanling male C57BL/6J x C3H/HeJ F1 mice," Cancer Research, vol. 47, no. 9, pp. 2275–2283, 1987.

161. W. Goggelmann and O. Schimmer, "Mutagenicity testing of β-asarone and commercial calamus drugs with Salmonella typhimurium," Mutation Research, vol. 121, no. 3-4, pp. 191–194, 1983.

162. G. Abel, "Chromosomenschadigende Wiruking von β-asarone in menschlichen Lymphocyten," Planta Medica, vol. 53, no. 3, pp. 251–253, 1987.

163. J. Lu and Y. Shi, "The bioactivity of essential oil from Ailanthus altissima Swingle (Sapindales: Simasoubaceae) bark on Lasioderma serricorne (Fabricus) (Coleoptera: Anobiidae)," Advanced Material Research, vol. 365, pp. 428–432, 2012.

164. D. Obeng-Ofori, "The use of botanicals by resource poor farmers in Africa and Asia for the protection of stored agricultural products," Stewart Postharvest Review, vol. 3, no. 6, article 10, 2007.

165. D. Obeng-Ofori, "Residual insecticides, inert dusts and botanicals for the protection of durable stored products against pest infestation in developing countries," in Stored Products Protection, M. O. Carvalho, P. G. Fields, C. S. Adler, et al., Eds., Proceedings of the 10th International Working Conference on Stored Product Protection, pp. 774–788, Julius-Kühn-Archiv 425, 2010.

Chapter 2

Microwave Heating as an Alternative Quarantine Method for Disinfestation of Stored Food Grains

Ipsita Das[1], Girish Kumar[1], and Narendra G. Shah[2]

[1]Department of Electrical Engineering, Indian Institute of Technology, Bombay, Powai, Mumbai 400076, India

[2]Centre for Technology Alternatives for Rural Areas, Indian Institute of Technology, Bombay, Powai, Mumbai 400076, India

ABSTRACT

Insects and pests constitute a major threat to food supplies all over the world. Some estimates put the loss of food grains because of infestation to about 40% of the world production. Contemporary disinfestation methods are chemical fumigation, ionizing radiation, controlled atmosphere, conventional hot air treatment, and dielectric heating, that is, radio frequency and microwave energy, and so forth. Though chemical fumigation is being used extensively in stored food grains, regulatory issues, insect resistance, and environmental concerns demand technically effective and environmentally sound quarantine methods. Recent studies have indicated that microwave treatment is a potential means of replacing other techniques because of selective heating, pollution free environment, equivalent or better quality retention, energy minimization, and so forth. The current paper reviews the recent advances in Microwave (MW) disinfestation of stored food products and its principle and experimental results from previous studies in order to establish the usefulness of this technology.

IMPORTANCE OF DISINFESTATION

Agricultural commodities produced on the fields have to undergo a series of operations such as harvesting, threshing, winnowing, bagging, transportation, storage, and processing before they reach the consumer, and there are appreciable losses in crop output at all these stages. Various estimates have been made to assess the postharvest food grain losses. The losses are caused either by environmental factors such as temperature, moisture, and type of storage structure or by biological agents, namely, insects, rodents, birds, and fungi. The major losses during production, storage and marketing of food grain are being attributed to infestation by insect pests, microbiological contamination, and physiological changes. Insect infestations can occur just prior to harvest or during storage or in-transit in a variety of carriers. The occurrence and numbers of stored grain insect pests are directly related to geographical and climatic conditions [1]. Almost all species have remarkably high rates of multiplication and, within one season, may destroy 10–15% of the grain and contaminate the rest with undesirable odors and flavors. Insect pests also play a pivotal role in transportation of storage fungi [2]. Therefore, preventing economic losses caused by stored-product insects is important from the field to the consumer's table [3]. The losses during storage are classified as quantity losses and quality losses. Quantity losses occur when the grain is consumed by insects, rodents, mites, birds and microorganisms. Quality losses are reflected as reduced economic value of the crop. The stored-grain insects affect not only the quantity of grain but also the quality of grain. Infestation causes decreased nutritional value, reduced seed germination, and lower economic value and also causes changes in chemical compositions such as increase in moisture, free fatty acid levels, nonprotein nitrogen content, and decrease in pH and protein contents in food grain [4]. They reduce the product quality directly by damage through feeding and indirectly by producing webbing and frass [5]. Also the grain quality has been found to decrease with time with increasing levels of infestation [6]. It is estimated that more than 20,000 species of field and storage pests destroy approximately one-third of the world's food production, valued annually at more than 100 billion dollar [7]. The quantitative and qualitative damage to stored grains and grain product from the insect pests may amount to 20–30% in the tropical zone and 5–10% in the temperate zone [8]. Food grain production in India has reached 250 million tons in the year 2010-2011, in which nearly 20–25% food grains are damaged by stored grain insect pests [7]. With population growth and the amount of cultivatable land shrinking, grain losses will continue to be a problem in the developing countries. Control of stored-product pests is one of the major tasks because the damage inflicted to foodstuff is irreversible. Also with progressive increase in the quantity of food grains and necessity for longer storage periods, these losses will escalate unless disinfestation measures are improved. International organizations such as FDA [9] and FGIS [10] have set tolerances and grade standards regulating the number of insects and insect fragments above specified tolerances to make the product illegal for human consumption. In developed countries even the mere presence of a few insects in a bulk, at densities of considerably less than one insect per kg grain can cause a serious loss in its market value. In some developed countries grain can be downgraded or rejected completely if even a single live insect is found [11]. The efficient control and removal of stored grain pests from food commodities have long been the goal of entomologists throughout the world. Various methods of insect control have been practiced to save the grain. In recent years, technology has made marked progress in the study of disinfestation of stored-grain insects and of finding improved ways to control them especially to detect latent forms of infestation. The implementation of an insect-disinfestation method requires detailed analysis of all the elements of an infestation problem: the insects, their age, species, and distribution and their survival and developmental rates under different environmental conditions [12]. Conventional chemicals, grain protectants, and fumigants are extensively used around the world to control insect pests in stored commodities because of low cost, fast processing, and easy application. Greater regulation and restriction of methyl bromide use will likely increase the cost of the fumigant, as well as reduce its availability [13]. With the concerns about health hazards of chemical pesticides and their resulting environmental pollution, there is interest in developing alternative, nonchemical process protocol to control insect pests while retaining acceptable product quality. These include conventional hot air or water heating, controlled atmosphere, and dielectric heating (radio frequency (RF) and microwave (MW)). Currently, the use of chemical fumigation remains widespread and the efficient use of RF and MW methods for disinfestation is still in research stage. The current paper reviews the various disinfestation methods of stored food grains with special emphasis on recent advances in microwave disinfestation of stored food grains. The principle of microwave disinfestation, experimental results of quality characteristics of microwave-treated grains, and the challenges of microwave disinfestation have also been described.

VARIOUS METHODS OF INSECT CONTROL

The control of stored-product insects is very important if the quality of foodstuff is to be maintained. The goal of the control measure is to render the habitat unsuitable for the growth and reproduction of stored-product insects. The five major potential quarantine treatment methods used for disinfestation of several insect pests for both the domestic and international markets are chemical fumigation, ionizing radiation, controlled atmosphere, conventional hot air/water heating, and dielectric heating using radio frequency (RF) and microwave (MW) energy which have been described in this paper.

Chemical Method

Since the 1950s, chemical insecticides have been used extensively in grain storage facilities to control stored-product insect pests due to low cost, fast speed in processing, and ease of use. Most postharvest pest management programs, therefore, rely heavily on fumigants. Currently, over 2.5 million tons of chemicals, worth over 30 billion US dollars, are applied to crops in the world [14]. The chemicals used to control insects in the bulk stored-grains are composed of two classes, namely, contact insecticides and fumigants. Contact insecticides, such as malathion, chlorpyrifos-methyl, or deltamethrin, are sprayed directly on grain or structures which provide protection from infestation for several months [15]. Fumigants are gaseous insecticides applied to control insects in grains that are inaccessible by contact insecticide. Some of the commonly used fumigants are methyl bromide (MeBr) and phosphine which rapidly kill all life stages of stored-product insects in food product [16]. Methyl bromide is now under the threat of withdrawal because it apparently depletes the Earth's ozone layer [17]. It has already been established that the use of MeBr has led to serious environmental damage and hazards to people's health. For these reasons, the Montreal Protocol constituted by the United Nations Environment Programme agreed on a phasing out of methyl bromide in the developed countries by 2005 and in the developing countries by 2015 [18]. Phosphine has been used as a replacement of methyl bromide for a long time [19]. Conventional use of phosphine has failed frequently to control insects [20]. Another limiting factor is that insects develop resistance to some particular chemical fumigants. Some of the contact insecticides have become ineffective because of wide-spread resistance in insect populations. A worldwide survey of stored-product insects revealed that 87% of 505 strains of the red flour beetle, Tribolium castaneum, collected from 78 countries were resistant to malathion [15]. Resistance to malathion is widespread in Canada, USA, and Australia [21] while resistance to phosphine is great in Australia and India, which may cause control failures [22]. The other major problem associated with the chemical methods is that, even if they are applied with care and in limited quantity, there is a possibility that these chemicals may remain in the food grains and have adverse effects on humans. Fumigation often only kills live larvae or adult insects but does not sterilize the eggs which are still alive in the grain kernels [23]. The consumption of organic products is also increasing each year. There is, therefore, interest in developing an alternative, nonchemical process method to control insect pests in food grains so as to minimize the environmental hazards associated with chemical insecticides while retaining acceptable product quality.

Ionizing Radiation

It is a process where infested food products are being exposed to ionizing radiation so as to sterilize, kill, or prevent emergence of insect pests by damaging their DNA. Three types of ionizing radiation used on foods are gamma rays from radioactive cobalt-60 and cesium-137, high energy electrons, and X-rays [24]. Irradiation with high energy electrons is usually safer and easier to work with because it can be turned on and off while an isotope is always radiating and humans must be shielded from it [25]. The ability of gamma rays to deeply penetrate pallet loads of food makes it one of the most commonly used in postharvest pest control. Sterilization of many species of insects can be accomplished at lower doses. Rusty grain beetles are sterilized at only 0.6 kGy but saw-toothed grain beetles and red flour beetles require a 2.0 kGy dose [26]. The grain mite, however, requires a much higher dose of 4.5 kGy. Though irradiation doses of 3–5 kGy were reported to be effective in controlling insects, they cause damage to product quality [26]. According to Hasan and Khan [27], high dosage of ionizing radiation has a risk of vitamin loss from the food product. The major drawback is substantial initial investment to establish the facility. In order to be economically feasible, the facility must remain in continuous operation. However, the seasonal nature of food produce prevents efficient use of facilities. Besides, this method has not received wide recognition because of high power consumption, large weight, and overall sizes of the installation. Consumers also have concerns over the disposal of radioactive wastes, the safety of the irradiation technology, and its effect on food [28]. Irradiation treatments often lead to live insects found by inspectors or consumers in the treated product because the applied doses do not immediately kill treated insects [29, 30]. Grains will continue to appear to be infested and a grain buyer cannot be certain that the insects are sterilized. Following irradiation with gamma rays at 0.5 kGy, complete insect mortality occurs in 14 days for rusty grain beetles, 28 days for red flour beetles, 70 days for saw-toothed grain beetles, and 200 days for grain mites. To date, irradiation is not accepted by the organic industry. Also approvals for irradiation of some selected food products (e.g., almond) have not been accepted by many countries, such as EU, Japan, and Taiwan [31].

Controlled Atmosphere Storage

Controlled atmosphere, a disinfestation technology wherein the normal composition of atmospheric air, that is, 21% O_2, 0.03% CO_2, and 78% nitrogen, is altered appropriately for disinfestation process [26]. An atmosphere containing more than 35% CO_2 known as carbon dioxide atmosphere and the atmosphere containing less than 1% oxygen, that is, low-oxygen atmosphere are lethal to insects. Controlled atmospheres are mainly based on the establishment of a low-oxygen environment which kills pests. The oxygen levels vary between 0% and 2%. It can be applied in airtight environments ranging from 1 M^3 to 1000 M^3 depending upon food commodity. Insects in all stages are eliminated because of the lack of oxygen which causes the insect to dry out and suffocate. This controlled atmosphere (CA) storage has been shown to be promising in creating lethal conditions for insects and fungi in stored food commodities. Annis and Morton [36] studied the effect of 15% to 100% CO_2 on developmental stages of insects in wheat at 25°C and 60% RH. They found that pupae were the most tolerant stages for all CO_2 concentrations and eggs were the only stages with 100% mortality at for less than 30 days. Gunasekaran and Rajendran [37] also found that the pupae stages were the most tolerant stages when exposed to different concentrations of CO_2. Controlled atmospheres are always being seen as average alternative such as longer treatment time, usability and availability, and being not suitable for dealing with high level of infestation.

Conventional Heat Treatment

Compared to the use of chemical methods, heat disinfestation has the potential for high market acceptance because of being residue-free. Most research focused on using hot air (80°C to 100°C) for disinfestation of food grains and showed satisfactory results. Thermal treatment methods using hot air/hot water alone or in combination with cold or controlled-atmosphere (CA) storage conditions have been investigated extensively for disinfestation of number of stored commodities [28, 38–55]. The hot air used to increase the temperature of the food product lies above the thermal limits of survival of the pest/insect. Heat disinfestation treatments are relatively easy to apply, leave no chemical residues, and may offer some fungicidal activity. There are a number of factors that affect the mortality of insects when exposed to hot air such as duration of exposure, temperature, species, and stage. Ideal temperatures for growth, reproduction, and movement for most stored product insects are between 25°C and 35°C [56]. Many insect larvae can bore into the center of fruits, nuts, seeds, or kernels, so the center of the commodity must be heated to lethal temperatures. Reported heating times for fruit center to reach the desired maximum temperatures range from 23 min for cherries to 6 h for apples [48, 49]. Conventional heating consists of heat transfer from the heating medium to the fruit surface and then conductive heat transfer from the surface to the center. A common difficulty with hot air or water heating methods is the slow rate of heat transfer resulting in long treatment time [57]. Prolonged heating is proved to be detrimental to the quality of final products such as peel browning, pitting, poor color development, and abnormal softening and may not be practical in industrial applications ([58], cited by [59]). There were reports of damage to grapefruits and mangos when exposed to forced air at 46°C for 3.75 h and 45°C for 1.8 h, respectively ([60, 61] cited by [48]). Flavor and appearance of air-heated grapefruits at 46°C for 3 h were inferior to those of unexposed fruits [62]. Surface browning in avocados was observed when heated with hot air at 43°C for 3.5 h [63] and in apples when exposed to hot water treatment at 46°C for 45 min [64]. A long exposure time requirement also causes alterations to flavor compounds [65]. The low heating rates also may increase the thermotolerance of the few insects [66–68] which was caused by the induction of heat shock proteins in insects during sublethal thermal conditions [69]. It is also important to determine the time-temperature combinations that result in 100% mortality for each insect over relatively large range of temperature. Sometimes temperature and time combinations required to kill the target insects may exceed those that reduce the crop nutrients, germination, or shelf life [70]. The disadvantages associated with conventional heat treatment method stimulated further studies on the possible use of dielectric heating for controlling stored-grain insects.

Dielectric Heating

Dielectric heating which covers both radio frequency (RF) and microwave (MW) has been investigated for insect control in foods [71]. Radio frequency (RF) heating is akin to microwave heating but utilizes another part of the electromagnetic spectrum. The frequency used for MW is 2450 MHz or 915 MHz while for RF the frequency is of 13, 27, or 40 MHz [48]. The effects of RF and MW energy are generally believed to be mainly thermal in nature [72]. Most agricultural products that considered dielectric material can store electric energy and convert electric energy into heat [49]. As the wavelength of RF heating frequencies is 22 to 360 times as more than that of the microwave frequencies, this allows RF energy to penetrate dielectric materials more deeply than microwaves. Many studies have explored the use of RF heating for control of insects in agricultural commodities [31, 47, 49, 73–82]. Researchers have reported the acceptable product quality after treating nuts and rice with RF energy to control insect infestation [48, 76–78, 83–85]. Recently, Wang et al. [81] studied postharvest disinfestation treatments for chickpeas and lentils using RF energy and also reported acceptable product quality. But the radio frequency heating is particularly useful when applied to institutional-sized packaged food products because of its deep penetration [86, 87].

Thermal treatment methods involving microwave radiation have extensively been investigated by several researchers as an alternative method of killing insects. Microwave quarantine method seems to have a great potential as an alternative method of killing insects in stored grain because of several advantages such as the control of all developmental stages of storage pests, having no chemical residues on the food product, having minimal impact on the environment, and providing rapid heating [28, 88–90]. Insects are unlikely to develop resistance to this treatment [91]. This electromagnetic energy (MW and RF) interacts directly with the product's interior to quickly raise the center temperature [47–49]. Microwave radiation not only kills insects by the dielectric heat induced within them but also affects the reproduction of survivors [92]. Microwave radiation, with good penetrability, can kill pests existing inside or outside grain kernels [93]. This paper brings the research initiatives especially on microwave (MW) heating for their potential use for disinfestation of stored food grains.

Microwave Heating

Microwaves are electromagnetic waves with frequencies ranging from about 300 MHz to 300 GHz and corresponding wavelengths from 1 to 0.001 m [94]. In electromagnetic spectrum, microwaves lie between radio frequencies and infrared radiation. Microwave heating is based on the transformation of alternating electromagnetic field energy into thermal energy by affecting polar molecules of a material. Many molecules in food (such as water and fat) are electric dipoles, meaning that they have a positive charge at one end and a negative charge at the other, and therefore they rotate as they try to align themselves with the alternating electric field induced by the microwave beam. The rapid movement of the bipolar molecules creates friction and results in heat dissipation in the material exposed to the microwave radiation. Microwave heating is most efficient on liquid water and much less on fats and sugars (which have less molecular dipole moment) [95]. The most important characteristic of microwave heating is volumetric heating which is different from conventional heating. The electromagnetic energy directly interacts with commodities to raise the interior temperature and significantly reduce treatment times as compared to conventional hot-water immersion and heated air methods. Conventional heating occurs by convection or conduction where heat must diffuse from the surface of the material. Volumetric heating means that materials can absorb microwave energy directly and internally and convert it into heat. The power generated in a material is proportional to the frequency of the source, the dielectric loss of the material, and the square of the field strength within it. The conversion of microwave energy to heat is expressed by the following equation [96]:

$$P = 2\pi E^2 f \varepsilon'' \varepsilon_0 V, \qquad (1)$$

Where P:power, W,E:the electric field strength, V/m,f:the frequency, Hz, ε_0 :the permittivity of free space, F/m, ε'' :the dielectric loss factor, and : volume of the material, m^3.

Dielectric properties of food depend on composition, temperature, bulk density, and microwave frequency. Since the influence of a dielectric depends on the amount of mass interacting with the electromagnetic fields, the mass per unit volume, or density, will also have an effect on the dielectric properties. This is especially notable with particulate dielectrics such as food grains. For granular and particulate materials, both the dielectric constant and loss factor tend to increase linearly with increasing bulk density of the materials especially in lower moisture materials such as cereal grains [97].

Microwave Disinfestation

The use of microwaves for disinfestation is based on the dielectric heating effect produced in grain, which is a relatively poor conductor of electricity. An attractive feature of the insect control using the microwave energy is that the insects are heated at a faster rate than the product they infest because of high moisture content of insects. So, it is possible to heat the insects to a lethal temperature because of their high moisture content while leaving the drier foodstuff unaffected or slightly warm [12]. Raising the temperature of infested materials by any means can be used to control insects if the infested product can tolerate the temperature levels that are necessary to kill the insects. There has been a lot of research on microwave disinfestation of cereals especially wheat [33, 98–104] and of some other food materials such as nuts [28], corn [105], pulses [106, 107], and cherries [89]. Exposure to microwave energy could cause physical injuries and reduced reproduction rates in surviving insects [75].

Feasibility of microwave disinfestation of insect pests has been explored by Andreuccetti et al. [74] for woodworms and by Halverson et al. [108] for wheat, maize, and flour weevils (cited by [89]). Nelson [75] reviewed the susceptibility of various stored grain insect species to radio frequency and microwave treatments. The use of microwave energy to control insects was initiated by Webber et al. [109] and the literature prior to 1980 has been reviewed by Tilton and Vardell [98]. Researchers have reported that microwave treatment is an attractive quarantine treatment. The major advantage of microwave heating is that it interacts directly with food grains and significantly reduce the amount of time required for food grain to reach the lethal temperature for insects as compared with conventional heating methods. Based on microwave heating theory and dissipated power calculations in (1), the absorption of microwave energy is proportional to the dielectric constant and dielectric loss factor of the materials. The dielectric properties of the materials depend on the frequency of the applied electric field and the temperature of the material [110]. If the material is hygroscopic, dielectric properties also depend on the amount of water in the material [111].

The first reported measurements of insect dielectric properties were for rice weevil and flour beetle at 40 MHz frequency. The dielectric constants were 6.6 and 7.8 for rice weevil and flour beetle, respectively, and loss factor was 2.2 for both species [112–114]. Nelson [110] presented data that confirms the dielectric loss factor of the insects was found to be much higher than that of stored-grain insects. The insect permittivity data at 25°C for 2.47 GHz frequency reported by Nelson et al. [32] are shown in Table 1. Alfaifi et al. [115] have reported that the loss factor of insect pests, such as Indian meal moth (Plodia interpunctella) and navel orangeworm (Amyelois transitella), is 26 to 36 times greater than that of dried fruits. Wang et al. [28, 48, 49] have indicated preferential heating of insects when walnut kernels and codling moth were subjected to radio frequency (27 MHz) and microwave frequency (2450 MHz) heating. Similar finding was observed by Ikediala et al. [116] and Raskovan et al. [117] for insect infested grain when subjected to radio frequency heating. They have reported that the dielectric constant and dielectric loss factor for granary weevil were greater than those of host grain in frequency band from 2 to 150 MHz. The reason attributed is that the higher the water content, the higher the values of dielectric properties of a material [118] which consist of dielectric constant (ε') and dielectric loss factor (ε'') described by the complex relative permittivity [115, 119]:

$$\varepsilon^* = \varepsilon' - j\varepsilon'', \qquad (2)$$

Where $j=(-1)^{0.5}$, ε' =real component, a measure of the ability of the material to store electromagnetic energy, and ε''= Imaginary component, a measure of the ability to dissipate electrical energy into heat.

Table 1: Dielectric properties of insects at 20–25°C ([32] cited by [33])

	Frequency (GHz)							
Adult insect species	**0.2**		**2.4**		**9.4**		**20**	
	ε'	ε''	ε'	ε''	ε'	ε''	ε'	ε''
S. oryzae	28	12	17	3	17	3	—	—
L. decemlineata	53	81	38	12	30	16	19	17

S. oryzae	42	28	32	9	25	12	18	13
S. oryzae	55	48	42	13	31	16	23	16
T. castaneum	61	56	47	15	34	19	25	19
O. surinamensis	70	68	53	17	40	21	28	22
R. dominica	63	55	43	15	34	19	25	18

ε' :dielectric constant, ε'' :dielectric loss.

Table 2 shows the dielectric permittivity of food when subjected to microwave heating. Hallman and Sharp [120] also summarized research on the application of radio frequency and microwave heat (electromagnetic energy) treatments to kill different pests on many postharvest food crops. Power level and exposure time of microwave have been identified as two important parameters to provide 100% insect mortality [101]. Microwave energies have been investigated for a number of food products other than cereal grains such as pulses, nuts, cherries, and dates. Recently, the effect of the microwave-heating method has been studied on disinfestations of the stored green gram seed [107]. The power level and exposure time were optimized based on insect mortality, color, and antinutrient factor with the value of power and exposure time of 800 W and 80 s, respectively. Singh et al. [106] have also studied the disinfestation of pulse beetle (adult stage) in chickpea, pigeon pea, and green gram as a function of exposure time and power level by exposing it continuously to microwave radiation (2450 MHz). The mortality of insect was found to increase with increase in microwave exposure time and power level or both. Though seed viability and germination of all these pulses were found to be affected, the cooking and milling characteristics were not affected by microwave exposure time and power level. Campana et al. [121] and Bhaskara Reddy et al. [122] that though eradication of the insects increased with the increase in microwave energy, but the seed viability, germination capacity, and seedling vigour decreased by exposure to microwave energy. Vadivambal et al. [104] and Vadivambal [33] had studied the mortality of different life stages of three common stored-grain insects, namely, Tribolium castaneum (Herbst),Cryptolestes ferrugineus (Stephens) and Sitophilus granarius (L.) in wheat, barley, and rye, respectively, using industrial microwave dryer operating at 2.45 GHz. Grain samples of 50 g at 14%, 16%, and 18% moisture content (wet basis) were exposed to four different power levels of 200, 300, 400, and 500 W for two exposure times of 28 and 56 s. Complete (100%) mortality was achieved for adults of all three insect species at 500 W for an exposure time of 28 s. The average temperature of wheat, barley, and rye at 500 W and 28 s was around 80°C, 71°C and 82°C, respectively. In all the cases, eggs were found to be most susceptible followed by larvae, and the least susceptible were the pupae and adults. Halverson et al. [103] had also reported that egg and young larva of all the three species were more susceptible than the pupa when microwave energy at 28 GHz frequency. The species tested were S. granarius, T. castaneum and R. dominica. Germination of seeds was lowered with an increase in power level or exposure time or both but there was no significant difference found in the quality characteristics of microwave-treated and control wheat, barley and rye except for a decrease in the flour yield. There were also reports by several researchers that no significant difference in the quality of grain protein, flour protein, flour yield, and loaf volume of sample was found when treated with microwave energy at which 100% mortality was obtained.

Table 2: Dielectric properties of selected food products at 20°C

Food product	Dielectric constant		Dielectric loss	
	915 MHz	2450 MHz	915 MHz	2450 MHz
Apple	57	54	8	10
Almond	2.1	—	2.6	—
Avocado	47	45	16	12
Banana	64	60	19	18
Carrot	59	56	18	15
Cucumber	71	69	11	12
Dates	12	—	5.7	—
Grape	69	65	15	17
Grapefruit	75	73	14	15
Lemon	73	71	15	14
Lime	72	70	18	15
Mango	64	61	12	14
Onion	61	64	12	14
Orange	73	69	14	16
Papaya	69	67	10	14
Peach	70	67	12	14
Pear	67	64	11	13
Potato	62	57	22	17
Radish	68	67	20	15

Strawberry	73	71	14	14
Walnut	3.2	—	6.4	—

Source: see [34].

Kaasova et al. [123] studied the chemical and biochemical changes during microwave treatment of wheat and reported that an improvement in the baking quality was found at higher energy doses and higher product temperatures. The decrease in germination capacity/seedling viability was related to the final temperature and the initial moisture content of the grains. Hamid and Boulanger [124] had found that the bread making quality of wheat was affected with increase in treatment temperature when exposed to power of 1.2 kW. They reported that 70% and 100% mortality were obtained when the wheat temperature was 55°C and 65°C, respectively, while, Locatelli and Traversa [125] have reported that temperature of grain has to reach 80°C for achieving complete mortality of insects infected using microwave. In fact, most infesting biological agents do not survive over a certain temperature called lethal temperature, generally between 55°C and 60°C, which can be rapidly reached through microwave irradiation [40]. There are three temperature zones for all insects: optimum, the zone at which highest rate of development can be achieved; suboptimum, a zone below or above optimum during which insects can complete their life cycle; and thirdly lethal zone, above or below suboptimum zones when insects get killed over period of time (Table 3). But the lethal temperatures to kill insects were found to vary considerably not only with species but also with the stage of development, temperature, and relative humidity. When developing effective treatment protocols, it is essential to determine which of the targeted insects is the most heat resistant, as well as the most heat resistant life stage for each species. To accomplish this, information is required on the minimum time-temperature combinations that result in 100% mortality for each insect over a relatively large range of temperature. The thermal death kinetics for fifth-instars of Indian meal moth, codling moth, and navel orange worm have been separately reported [76, 77, 126–129]. The authors of [130] recommended 15 min exposures to microwaves, which increased temperature to 60°C to kill all insects in confectionery walnuts without adversely affecting flavor.

Table 3: Response of stored-product insects to temperature [35]

Temperature (°C)	Zone	**Effect**
50–60	Lethal	Death in minutes
45		Death in hours
35	Suboptimum	Development stops
33–35		Development slows
25–33	Optimum	Maximum rate of development
13–25	Suboptimum	Development slows
13–20		Development stops
5	Lethal	Death in days
−10 to −5		Death in weeks to month
−25 to −15		Death in minutes, insects freeze

Halverson et al. [103] conducted experiments to determine life stages of the insect that is most susceptible to microwave energy at 28 GHz frequency. The life stages tested were egg, young larva, and pupa. Their results suggested that egg and young larva of all were always more susceptible than the pupa. Zouba et al. [131] had employed a research-scale microwave unit to investigate the mortality of insect in date and the thermal impact on the date quality. The heating characteristics of dates are highly influenced by initial moisture content. Date having various moisture contents gets heated up in a very heterogeneous manner and the energy of the microwaves seems to be preferentially absorbed by soft dates. Similar results were reported during insect control of walnuts using radio frequency treatments. The insect mortality is achieved in less than 90 s during microwave treatment without altering date quality.

Challenges in Microwave Disinfestation

Although microwaves have potential for disinfestation applications in the grain industry, they have not been used widely due to their adverse effects on various quality parameters. The major problems associated with microwave heating are the nonuniform temperature distribution and thereby incomplete kill of microbes [132–137]. Hot spots, produced on products due to the nonuniform heating pattern of microwaves, are one of the important factors for the quality degradation of products during microwave treatment. A hot spot can be defined as a local area of very high temperature that results from the temperature dependence of material properties. Although hot spots are desirable to increase insect mortality, detrimental effects on the medium are not. Manickavasagan et al. [138] had determined the germination percentage of microwave-treated wheat samples collected from hot-spot and normal heating zones. The wheat samples having four different moisture levels (12%, 15%, 18%, and 21% wb) were subjected to different microwave treatments (100, 200, 300, 400, and 500 watt with exposure times of 28 and 56 s) in continuous industrial microwave dryer (2450 MHz). The germination percentage of the sample in the hot-spot region was significantly lower than that of the normal heating region for all moisture and power levels. Apart from germination, the other quality parameters were also found to get affected more in the hot-spot zone than the remaining bulk grain. Moreover, the generation of "hot spots" during industrial microwave heating has recently become of great concern to scientists involved in microwave research [139, 140]. The other major drawback in microwave disinfestation is the poor penetrating power of microwaves. The microwave's intensity diminishes with increased penetration. It has been reported that microwave treatment of bulk grain is not feasible when the depth is greater than 4 inch ([92] cited by [33]). Due to the limited penetration of microwave energy into foodstuff mass, it seems likely that employment of microwave radiation alone could not be considered as a promising insect control measure under field condition. Disinfestations of stored products using microwaves energy coupled with other modes of treatment can be an alternative measure in killing insects effectively, but little work has been reported on combined application. The mechanisms involved in the lethal action of microwave radiation are already understood. Microwave radiation has effects on insects such as reduction of reproductive rate, losing body weight, and malformation [75]. However, application of microwaves radiation could be limited due to insufficient penetration depth. Songping et al., [140] reported that microwaves attenuate exponentially in penetration to foodstuffs.

Combined application of microwaves with hot-air treatment/cold storage/gamma radiation could be considered as a potential measure which can help reduce stored-product insects population. The combined impact of microwave radiation and cold storage on adults was evaluated by several researchers. Ayvaz and Karabörklü [141] had reported that there is a decrease in reproductive ability and number of living adults depending on the length of the cold storage period. In general, the reduction of temperature in the environment stresses the insect, thereby, making it more susceptible to other control measures [48, 89]. The major advantage is that the cold storage can easily be coupled with microwave radiation for pest control measure. There was sufficient indication that longer microwaves energy exposure and cold storage duration could achieve better kill than shorter ones of similar power level. When the insects such as Tribolium castaneum H. and Sitophilus oryzae L. were exposed to 2450 MHz at power level of 100 W for exposure time 10 min, continuously and intermittently, the highest rate of mortality was achieved for intermittent exposure time of 10 min and 72 h of cold storage duration [142]. Valizadegan et al. [143] have also evaluated the impact of microwave radiation in conjunction with cold storage on adult insects (saw-toothed grain beetle and cigarette beetle) under laboratory conditions. The insects were exposed to 2450 MHZ at five different power levels of 0, 100, 200, 300, and 400 W for five exposure times of 0, 3, 6, 9, and 12 min. The complete control was achieved at 400 W power levels for exposure time of 12 min and 72 h cold storage period. Similar results, that is, high mortality rate, were also reported by Nasab et al. [5] when Indian meal moth eggs were exposed to power levels of 100, 300, and 500 W with 2, 4, 6, 8, and 10 min exposure duration along and then kept in cold storage conditions (°C) for 24 and 48 h. Combinations of microwave radiation and cold storage were found to be compatible and synergistic which provide an effective and environmentally friendly disinfestation treatment technique.

Researchers have also tried the microwave heating along with gamma irradiation to control insects in stored cereals and cereals products. El-Naggar and Mikhaiel [144] evaluated the biochemical analyses on the samples of wheat grain and flour subjected to combined microwave and gamma irradiation treatment where high mortality was obtained. There were no detectable changes in the quality of protein, fat, fiber, carbohydrates, or ash. Amjad and Akbar Anjum [145] stated that when onion seeds (Allium cepa L.) were exposed to various doses of gamma radiation, that is, 0, 20, 40, 80, and 100 krad, there was no significant effect on seed viability except at the highest dose (100 krad) which resulted in reduced viability. But several other researchers reported that microwaves are not suitable for the drying of wheat which is to be used as seeds, even using low power levels, unless some provisions are made to ensure uniform heating [124, 146–148]. They have concluded that bread making quality of wheat and maize get adversely affected by exposure to microwave radiation. Though microwave heating helped in reducing the power consumption in wheat milling process, the textural characteristics of the final products are made from the microwave-treated flour were found not to be acceptable. The viscosity of the flour decreased with increased exposure time because of the alteration in structure of starch and protein when exposed to microwave energy. Campana et al. [121] had also stated that the protein content was not affected but the functionality of gluten altered gradually with increasing time of exposure. The change in the functionality of gluten is because of the absence of elasticity and stretchability of the dough [121, 147].

Several researchers concluded that germination capacity was affected by exposure to microwave energy. The seed viability and germination of chickpea, pigeon pea, and green gram were also reported to be affected by microwave exposure time and power level [106]. Vadivambal et al. [149] found that microwave energy at 2450 MHz with similar heating mechanism to RF treatments might control storage insects in barley and rye but resulted in poor germination due to high sample temperatures. The decrease in germination capacity was related to the final temperature and the initial moisture content of the grains. Nelson and Stetson [150] studied the dielectric heating treatments of rice weevils in wheat at 39 and 2450 MHz and showed that the lower frequency was much more effective in killing the insects. Wang et al. [86] had concluded that differential heating of insects in walnuts does occur at 27 MHz but not at 915 MHz based on the experiments carried out on dried nuts and fruits by exposure to microwave and radio frequencies range.

Bhaskara Reddy et al. [122] studied the effect of microwave treatment on quality of wheat seeds infected withFusarium graminearum (Schwabe). Their results showed that though mortality of insect increased with the microwave power level, but the seed viability and seedling vigor decreased accordingly. It has been concluded that microwave drying of wheat would not be suitable where the final products made out of flour required soft textural characteristics. Another issue with the microwave heating is the large number of factors that affect the microwave heat transfer behavior such as the thickness, the geometry, and the dielectric properties of the food. The heat capacity and the dielectric properties (dielectric constant, loss factor) change with the moisture content and temperature which complicates the microwave heating process. Cost estimates for microwave and radio frequency insect control are estimated to be around three to five times more expensive than chemical control.

CONCLUSIONS

Insects cause considerable damage to food grains with weight and nutritional losses reducing yields and market values. Postharvest phytosanitary treatments are often required to completely control insect pests before the products are moved through marketing channels to areas where the pests do not occur. Several methods have been suggested to control insect pests in agricultural commodities, including chemical fumigation, thermal treatment, ionizing radiation, cold storage, controlled atmospheres, dielectric heating, and combination treatments. Current technologies involve the use of toxic chemicals which is neither consumer friendly nor environmentally friendly and conventional thermal methods are either undesirable or cause loss of volatile components, browning, and texture change. To date, irradiation is not accepted by the organic industry. Based on the results of the microwave disinfestation studies already conducted, it is considered as safe and competitive alternative method to other quarantine methods and can avoid problems of food safety and environmental pollution. The study conducted on different food products infested with major insects says that complete mortality, that is, 100%, could be achieved using microwave energy. Although microwaves have potential for disinfesting the food products, they have not been used widely due to their adverse effects on various quality parameters. Nonuniformity of heating is one of the important factors that cause quality deterioration of food product. Though several practical means can be used to minimize non-uniform heating such as adding forced hot air to the product surface or sample movement/rotation or mixing during treatment, and so forth. Also if the microwave disinfestation is made economical, it may serve as a safe and effective alternate method of insect control. The most important factor in the development of an acceptable insect control method using microwave energy is to identify a balance between minimized thermal impact on the product quality and complete killing of the insect population. To achieve a balance between complete eradication of the insects and to maintain the product quality, further research needs to be done on large scale tests, with infested product to confirm the treatment efficacy and product quality after extended storage before this technology would provide an acceptable process for disinfestation.

ACKNOWLEDGMENT

The authors acknowledge the Department of Science and Technology (DST), India, for providing financial support for this study.

REFERENCES

1. S. Lal and B. P. Srivastava, "Insect pests of stored wheat of Madhya Pradesh (India)," Journal of Entomological Research, vol. 9, pp. 141–148, 1985.
2. A. K. Sinha and K. K. Sinha, "Insect pests, Aspergillus flavus and aflatoxin contamination in stored wheat: a survey at North Bihar (India)," Journal of Stored Products Research, vol. 26, no. 4, pp. 223–226, 1990.
3. B. Subramanyam and D. W. Hagstrum, Alternatives to Pesticides in STored-Product IPM, Kluwer Academic Publishers, Norwell, Mass, USA, 2000.
4. R. I. Sánchez-MariñEz, M. O. Cortez-Rocha, F. Ortega-Dorame, M. Morales-Valdes, and M. I. Silveira, "End-use quality of flour from Rhyzopertha dominica infested wheat," Cereal Chemistry, vol. 74, no. 4, pp. 481–483, 1997.
5. F. S. Nasab, A. A. Pourmirza, and A. H. Zade, "The effect of microwave radiation with cold storage on the mortality of Indian meal moth (Plodia Interpunctella Hub.) Eggs," Pakistan Journal of Entomology, vol. 31, no. 2, pp. 111–115, 2009.
6. J. P. Edwards, J. E. Short, and L. Abraham, "Large-scale evaluation of the insect juvenile hormone analogue fenoxycarb as a long-term protectant of stored wheat," Journal of Stored Products Research, vol. 27, no. 1, pp. 31–39, 1991.
7. Y. Rajashekar, N. Bakthavatsalam, and T. Shivanandappa, "Review article on botanicals as grain protectants," vol. 2012, Article ID 646740, 13 pages, 2012.

8. S. Rajendran and V. Sriranjini, "Plant products as fumigants for stored-product insect control," Journal of Stored Products Research, vol. 44, no. 2, pp. 126–135, 2008.
9. FDA: Food Drug Administration, Wheat Flour-Adulteration with Insect Fragments and Rodent Hairs (CPG, 7104. 06). Compliance Policy Guides Manual, chapter 5, Foods, color and cosmetics, sub chapter 578-processed grain, 1997.
10. FGIS: Federal Grain Inspection Service, Official United States standards for grain subpart m. United States standards for wheat, 1999.
11. D. B. Pinniger, M. R. Stubbs, and J. Chambers, "The evaluation of some food attractants for the detection of Oryzaephilus surinamensis (L.) and other storage pests," in Proceedings of the 3rd International Working Conference on Stored Product Entomology, pp. 640–650, Kansas, USA, October 1984.
12. R. Vadivambal, D. S. Jayas, and N. D. White, "Disinfestation of life stages of Tribolium castaneum in wheat using microwave energy," in Proceedings of the CSBE/SCGAB, 2006 Annual Conference Edmonton Alberta, 2006, Paper No. 06-120S.
13. E. Mitcham, "Quarantine issues in 2000," Acta Horticulture, vol. 553, pp. 451–455, 2001.
14. J. Gannage, "Pesticides and human health," 2000, http://www.herbs2000.com/articles/pesticides.htm.
15. R. N. Sinha and F. L. Watters, Insect Pests of Flour Mills, Grain Elevators, and Feed Mills and Their Control, Agriculture Canada, Winnipeg, Canada, 1985.
16. E. J. Bond, Manual of Fumigation For Insect Control, Food and Agricultural Organization, Rome, Italy, 1984.
17. J. G. Leesch, G. F. Knapp, and B. E. Mackey, "Methyl bromide adsorption on activated carbon to control emissions from commodity fumigations," Journal of Stored Products Research, vol. 36, no. 1, pp. 65–74, 2000.
18. UNEP (United Nations Environment Programme), Report of the 9th Meeting of the Parties To the Montreal ProTocol on Substances That Deplete the Ozone Layer, UNEP/OzL.Pro. 9/12, Montreal, Canada, 1997.

19. S. Rajendran and N. Muralidharan, "Performance of phosphine in fumigation of bagged paddy rice in indoor and outdoor stores," Journal of Stored Products Research, vol. 37, no. 4, pp. 351–358, 2001.
20. C. H. Bell and S. M. Wilson, "Phosphine tolerance and resistance in Trogoderma granarium Everts (Coleoptera: Dermestidae)," Journal of Stored Products Research, vol. 31, no. 3, pp. 199–205, 1995.
21. B. H. Subramanyam and D. Hagstrum, "Resistance measurement and management," in Integrated Management of Insects in Stored Products, B. Subramanyam and D. Hagstrum, Eds., pp. 331–398, Marcel Dekker, New York, NY, USA, 1995.
22. P. Collins, G. Daglish, H. Pavic, T. Lambkin, R. Kopittke, and B. Bridgeman, "Combating strong resistance to phosphine in stored grain pests in Australia," in Proceedings of the 2nd Australian Postharvest Technical Conference, E. J. Wright, H. J. Banks, and E. Highley, Eds., pp. 109–112, Adelaide, South Australia, 2000.
23. S. J. Langlinais, "Economics of microwave treated rice for controlling weevils," ASAE Paper 893544, ASABE, St. Joseph, Mich, USA, 1989.
24. E. Shadia and S. E. Abd El-Aziz, "Control strategies of stored product pests," Journal of Entomology, vol. 8, no. 2, pp. 101–122, 2011.
25. P. G. Fields and W. E. Muir, "Physical control," in Integrated Management of Insects in Stored Products, Marcel Dekker, New York, NY, USA, 1996.
26. J. Banks and J. B. Fields, "Physical methods for insect control in stored-grain ecosystems," in Stored-Grain Ecosystem, D. S. Jayas, N. D. G. White, and W. E. Muir, Eds., pp. 353–410, Marcel Dekker, New York, NY, USA, 1995.
27. M. Hasan and A. R. Khan, "Control of stored-product pests by irradiation," Integrated Pest Management Reviews, vol. 3, no. 1, pp. 15–29, 1998.
28. S. Wang and J. Tang, "Radio frequency and microwave alternative treatments for insect control in nuts: A review," International Agricultural Engineering Journal, vol. 10, no. 3-4, pp. 105–120, 2001.

29. G. Saour and H. Makee, "Susceptibility of potato tuber moth (Lepidoptera: Gelechiidae) to postharvest gamma irradiation," Journal of Economic Entomology, vol. 97, no. 2, pp. 711–714, 2004.
30. N. W. Heather, G. Hallman, and N. Heather, Pest Management and Phytosanitary Trade Barriers, CABi Publishing, Cambridge, Mass, USA, 2008.
31. M. Gao, J. Tang, Y. Wang, J. Powers, and S. Wang, "Almond quality as influenced by radio frequency heat treatments for disinfestation," Postharvest Biology and Technology, vol. 58, no. 3, pp. 225–231, 2010.
32. S. O. Nelson, P. G. Hartley, and K. C. Lawrence, "RF and microwave dielectric properties of stored-grain insects and their implications for potential insect control," Transactions of the American Society of Agricultural Engineers, vol. 41, no. 3, pp. 685–692, 1998.
33. R. Vadivambal, Disinfestation of Stored Grain Insects Using Microwave Energy [Ph.D. thesis], University of Manitoba, Winnipeg, Canada, 2009.
34. M. S. Venkatesh and G. S. V. Raghavan, "An overview of microwave processing and dielectric properties of agri-food materials," Biosystems Engineering, vol. 88, no. 1, pp. 1–18, 2004.
35. M. P. Fakude, "Eradication of storage insect pests in maize using microwave energy and the effects of the latter on grain quality," 2007, http://upetd.up.ac.za/thesis/available/etd-01292009-131525/unrestricted/dissertation.pdf.
36. P. C. Annis and R. Morton, "The acute mortality effects of carbon dioxide on various life stages of Sitophilus oryzae," Journal of Stored Products Research, vol. 33, no. 2, pp. 115–124, 1997.
37. N. Gunasekaran and S. Rajendran, "Toxicity of carbon dioxide to drugstore beetle Stegobium paniceumand cigarette beetle Lasioderma serricorne," Journal of Stored Products Research, vol. 41, no. 3, pp. 283–294, 2005. Vie
38. D. E. Evans, "Some biological and physical constraints to the use of heat and cold for disinfesting and preserving stored products," in Proceedings of the 4th International Working Conference of

Stored Product Protection, E. Donahaye and S. Navarro, Eds., pp. 149–164, Tel-Aviv, Israel, 1986.

39. J. L. Sharp, J. J. Gaffney, J. I. Moss, and W. P. Gould, "Hot-air treatment device for quarantine research,"Journal of Economics Entomology, vol. 84, pp. 520–527, 1991.

40. P. G. Fields, "The control of stored-product insects and mites with extreme temperatures," Journal of Stored Products Research, vol. 28, no. 2, pp. 89–118, 1992.

41. V. Y. Yokoyama, G. T. Miller, and R. V. Dowell, "Response of codling Moth (Lepidoptera: Tortricidae) to high temperature, a potential quarantine treatment for exported commodities," Journal of Economics Entomology, vol. 84, pp. 528–531, 1991.

42. L. G. Neven, "Combined heat treatments and cold storage effects on mortality of fifth-instar codling moth (Lepidoptera: Tortricidae)," Journal of Economic Entomology, vol. 87, no. 5, pp. 1262–1265, 1994.

43. L. G. Neven and L. M. Rehfield, "Comparison of Pre-Storage Heat Treatments on Fifth-Instar Codling Moth (Lepidoptera: Tortricidae) Mortality," Journal of Economic Entomology, vol. 88, pp. 1371–1375, 1995.

44. L. J. Mason and C. A. Strait, "Stored product integrated pest management with extreme temperatures," in Temperature Sensitivity in Insects and Application in Integrated Pest Management, G. J. Hallman and D. L. Denlinger, Eds., pp. 141–177, Westview Press, Boulder, Colo, USA, 1998.

45. A. K. Dowdy, "Heat sterilization as an alternative to methyl bromide fumigation in cereal processing plants," in Proceedings of 7th International Working Conference for Stored Product Protection, Z. Jin, Q. Liang, Y. Liang, X. Tan, and L. Guan, Eds., pp. 1089–1095, Beijing, China, 1999.

46. C. S. Burks, J. A. Johnson, D. E. Maier, and J. W. Heaps, "Temperature.," in Alternatives to Pesticides in stored Product IPM, B. Subramanyam and D. W. Hagstrum, Eds., pp. 73–104, Kluwer Academic Publishers, Boston, Mass, USA, 2000.

47. J. Tang, J. N. Ikediala, S. Wang, J. D. Hansen, and R. P. Cavalieri, "High-temperature-short-time thermal quarantine methods," Postharvest Biology and Technology, vol. 21, no. 1, pp. 129–145, 2000.

48. S. Wang, J. N. Ikediala, J. Tang et al., "Radio frequency treatments to control codling moth in in-shell walnuts," Postharvest Biology and Technology, vol. 22, no. 1, pp. 29–38, 2001.
49. S. Wang, J. Tang, and R. P. Cavalieri, "Modeling fruit internal heating rates for hot air and hot water treatments," Postharvest Biology and Technology, vol. 22, no. 3, pp. 257–270, 2001.
50. S. Lurie, T. Jemric, A. Weksler, R. Akiva, and Y. Gazit, "Heat treatment of 'Oroblanco' citrus fruit to control insect infestation," Postharvest Biology and Technology, vol. 34, no. 3, pp. 321–329, 2004.
51. M. Schirra, M. Mulas, A. Fadda, I. Mignani, and S. Lurie, "Chemical and quality traits of "Olinda" and "Campbell" oranges after heat treatment at 44 or 46°C for fruit fly disinfestation," LWT—Food Science and Technology, vol. 38, no. 5, pp. 519–527, 2005.
52. O. Dosland, B. Subramanyam, K. Sheppard, and R. Mahroof, "Temperature modification for insect control," in Insect Management For Food Storage and Processing, J. Heaps, Ed., pp. 89–103, American Association of Cereal Chemists, St. Paul, Minn, USA, 2005.
53. S. J. Beckett, P. G. Fields, and B. Subramanyam, "Disinfestation of stored products and associated structures using heat," in Heat Treatments For Postharvest Pest Control: Theory and Practice, J. Tang, E. Mitcham, S. Wang, and S. Lurie, Eds., CAB International, Wallingford, UK, 2006.
54. O. Dosland, B. Subramanyam, K. Sheppard, and R. Mahroof, "Temperature modification for insect control," in Insect Management For Food Storage and Processing, J. Heaps, Ed., pp. 89–103, American Association for Clinical Chemistry, St. Paul, Minn, USA, 2nd edition, 2006.
55. E. J. Mitcham, S. Zhou, and V. Bikoba, "Controlled atmospheres for quarantine control of three pests of table grape," Journal of Economic Entomology, vol. 90, no. 5, pp. 1360–1370, 1997.
56. R. W. Howe, "A summary of estimates of optimal and minimal conditions for population increase of some stored products insects," Journal of Stored Products Research, vol. 1, no. 2, pp. 177–184, 1965.

57. J. Hansen, "Heating curve models of quarantine treatments against insect pests," Journal of Economics Entomology, vol. 85, pp. 1846–1854, 1992.
58. S. Lurie, "Postharvest heat treatments," Postharvest Biology and Technology, vol. 14, no. 3, pp. 257–269, 1998.
59. S. Wang, X. Yin, J. Tang, and J. D. Hansen, "Thermal resistance of different life stages of codling moth (Lepidoptera: Tortricidae)," Journal of Stored Products Research, vol. 40, no. 5, pp. 565–574, 2004.
60. G. J. Hallman, J. J. Gaffney, and J. L. Sharp, "Vapor heat treatment for grapefruit infested with Caribbean fruit fly (Diptera: Tephritidae)," Journal of Economic Entomology, vol. 83, no. 4, pp. 1475–1478, 1990.
61. D. Ortega-Zaleta and E. M. Yahia, "Tolerance and quality of mango fruit exposed to controlled atmospheres at high temperatures," Postharvest Biology and Technology, vol. 20, no. 2, pp. 195–201, 2000.
62. K. C. Shellie, R. L. Mangan, and S. J. Ingle, "Tolerance of grapefruit and mexican fruit fly larvae to heated controlled atmospheres," Postharvest Biology and Technology, vol. 10, no. 2, pp. 179–186, 1997.
63. E. L. Kerbel, G. Mitchell, and G. Mayer, "Effect of postharvest heat treatment for insect control on the quality and market life of avocados," Horticultural Science, vol. 22, pp. 92–94, 1987.
64. K. J. Smith and M. Lay-Yee, "Response of "Royal Gala" apples to hot water treatment for insect control,"Postharvest Biology and Technology, vol. 19, no. 2, pp. 111–122, 2000.
65. D. M. Obenland, M. L. Arpaia, R. K. Austin, and B. E. MacKey, "High-temperature forced-air treatment alters the quantity of flavor-related, volatile constituents present in navel and valencia oranges," Journal of Agricultural and Food Chemistry, vol. 47, no. 12, pp. 5184–5188, 1999.
66. D. B. Thomas and K. C. Shellie, "Heating rate and induced thermotolerance in Mexican fruit fly (diptera: tephritidae) larvae, a quarantine pest of citrus and mangoes," Journal of Economic Entomology, vol. 93, no. 4, pp. 1373–1379, 2000.

67. S. J. Beckett and R. Morton, "Mortality of Rhyzopertha dominica (F.) (Coleoptera: Bostrychidae) at grain temperatures ranging from 50°C to 60°C obtained at different rates of heating in a spouted bed," Journal of Stored Products Research, vol. 39, no. 3, pp. 313–332, 2003.

68. S. Wang and J. Tang, "Heating uniformity and differential heating of insects in almonds in radio frequency systems," Paper 076019, American Society of Agricultural and Biological Engineers, St. Joseph, Mich, USA, 2007.

69. X. Yin, S. Wang, J. Tang, J. D. Hansen, and S. Lurie, "Thermal conditioning of fifth-instar Cydia pomonella (Lepidoptera: Tortricidae) affects HSP70 accumulation and insect mortality," Physiological Entomology, vol. 31, no. 3, pp. 241–247, 2006.

70. D. E. Evans, G. R. Thorpe, and T. Dermott, "The disinfestation of wheat in a continuous-flow fluidized bed," Journal of Stored Products Research, vol. 19, no. 3, pp. 125–137, 1983.

71. Y. C. Fu, "Fundamentals and industrial applications of microwave and radio frequency in food processing," in Food Processing: Principles and Applications, J. S. Smith and Y. H. Hui, Eds., Blackwell, Iowa City, IA, Iowa, USA, 2004.

72. M. A. Stuchly, "Health effects of exposure to electromagnetic fields," IEEE Aerospace Applications Conf Proceeding, vol. 1, pp. 351–368, 1995.

73. S. O. Nelson and J. A. Payne, "RF dielectric heating for pecan weevil control," Transactions of the American Society of Agricultural Engineers, vol. 31, pp. 456–458, 1982.

74. D. Andreuccetti, M. Bini, A. Ignesti, A. Gambetta, and R. Olmi, "Microwave destruction of woodworms," Journal of Microwave Power and Electromagnetic Energy, vol. 29, no. 3, pp. 153–160, 1994.

75. S. O. Nelson, "Review and assessment of radio-frequency and microwave energy for stored-grain insect control," Transactions of the American Society of Agricultural Engineers, vol. 39, no. 4, pp. 1475–1484, 1996.

76. S. Wang, J. N. Ikediala, J. Tang, and J. D. Hansen, "Thermal death kinetics and heating rate effects for fifth-instar Cydia pomonella (L.) (Lepidoptera: Tortricidae)," Journal of Stored Products Research, vol. 38, no. 5, pp. 441–453, 2002.

77. S. Wang, J. Tang, J. A. Johnson, and J. D. Hansen, "Thermal-death kinetics of fifth-instar Amyelois transitella (Walker) (Lepidoptera: Pyralidae)," Journal of Stored Products Research, vol. 38, no. 5, pp. 427–440, 2002.
78. S. Wang, J. Tang, J. A. Johnson et al., "Process protocols based on radio frequency energy to control field and storage pests in in-shell walnuts," Postharvest Biology and Technology, vol. 26, no. 3, pp. 265–273, 2002.
79. E. J. Mitcham, R. H. Veltman, X. Feng et al., "Application of radio frequency treatments to control insects in in-shell walnuts," Postharvest Biology and Technology, vol. 33, no. 1, pp. 93–100, 2004.
80. F. Marra, L. Zhang, and J. G. Lyng, "Radio frequency treatment of foods: review of recent advances,"Journal of Food Engineering, vol. 91, no. 4, pp. 497–508, 2009.
81. S. Wang, G. Tiwari, S. Jiao, J. A. Johnson, and J. Tang, "Developing postharvest disinfestations treatments for legumes using radio frequency energy," Biosystems Engineering, vol. 105, no. 3, pp. 341–349, 2010.
82. S. Jiao, J. A. Johnson, J. K. Fellman et al., "Evaluating the storage environment in hypobaric chambers used for disinfesting fresh fruits," Biosystems Engineering, vol. 111, no. 3, pp. 271–279, 2012.
83. S. Wang, M. Monzon, J. A. Johnson, E. J. Mitcham, and J. Tang, "Industrial-scale radio frequency treatments for insect control in walnuts. I: Heating uniformity and energy efficiency," Postharvest Biology and Technology, vol. 45, no. 2, pp. 240–246, 2007.
84. S. Wang, M. Monzon, J. A. Johnson, E. J. Mitcham, and J. Tang, "Industrial-scale radio frequency treatments for insect control in walnuts. II: Insect mortality and product quality," Postharvest Biology and Technology, vol. 45, no. 2, pp. 247–253, 2007.
85. M. C. Lagunas-Solar, Z. Pan, N. X. Zeng, T. D. Truong, R. Khir, and K. S. P. Amaratunga, "Application of radiofrequency power for non-chemical disinfestation of rough rice with full retention of quality attributes," Applied Engineering in Agriculture, vol. 23, no. 5, pp. 647–654, 2007.
86. S. Wang, J. Tang, R. P. Cavalieri, and D. C. Davis, "Differential heating of insects in dried nuts and fruits associated with radio

frequency and microwave treatments," Transactions of the American Society of Agricultural Engineers, vol. 46, no. 4, pp. 1175–1182, 2003.

87. S. Wang, J. Tang, J. A. Johnson et al., "Dielectric properties of fruits and insect pests as related to radio frequency and microwave treatments," Biosystems Engineering, vol. 85, no. 2, pp. 201–212, 2003.

88. S. L. Halverson, R. Plarre, T. S. Bigelow, and K. Lieber, "Recent advance in the use of EHF energy for the control of insect in stored products," in Proceedings of the ASAE Annual International Meeting, Orlando, Fla, USA, 1998, Paper No. 986052.

89. J. N. Ikediala, J. Tang, L. G. Neven, and S. R. Drake, "Quarantine treatment of cherries using 915 MHz microwaves: temperature mapping, codling moth mortality and fruit quality," Postharvest Biology and Technology, vol. 16, no. 2, pp. 127–1, 1999.

90. O. A. Karabulut and N. Baykal, "Evaluation of the use of microwave power for the control of postharvest diseases of peaches," Postharvest Biology and Technology, vol. 26, no. 2, pp. 237–240, 2002.

91. F. L. Watters, "Microwave radiation for control of Tribolium confusum in wheat and flour," Journal of Stored Products Research, vol. 12, no. 1, pp. 19–25, 1976.

92. M. A. K. Hamid, C. S. Kashyap, and R. V. Cauwenberghe, "Control of grain insects by microwave power," Journal of Microwave Power, vol. 3, no. 3, pp. 126–135, 1968.

93. S. L. Halverson, T. W. Phillips, T. S. Bigelow, G. N. Mbata, and M. E. Payton, "The control of various species of stored-product insects with EHF energy," in Proceeding of the Annual International Research Conference on Methyl Bromide Alternatives and Emissions Reductions, pp. 54-1–54-4, 1999.

94. R. V. Decareau, Microwaves in the Food Processing Industry, Academic Press, Natick, Mass, USA, 1985.

95. P. P. Sutar and S. Prasad, "Microwave drying technology-recent developments and R&D needs in India," in proceedings of the 42nd ISAE Annual Convention, pp. 1–3, Kolkata, India, February 2008.

96. H. Linn and M. Moller, "Microwave heating," in Proceedings of the Thermprocess Symposium, pp. 16–21, Dusseldorf, Germany, June 2003.
97. S. O. Nelson and S. Trabelsi, "Factors influencing the dielectric properties of agricultural and food products," Journal of Microwave Power and Electromagnetic Energy, vol. 46, no. 2, pp. 93–107, 2012.
98. E. W. Tilton and H. H. Vardell, "Combination of microwaves and partial vacuum for control of four stored-product insects in stored grain," Georgia Entomological Society, vol. 17, pp. 96–106, 1982.
99. E. W. Tilton and J. H. Brower, "Ionizing radiation for insect control in grain and grain products," Cereal Foods World, vol. 32, no. 4, pp. 330–335, 1987.
100. N. Shayesteh and N. N. Barthakur, "Mortality and behaviour of two stored-product insect species during microwave irradiation," Journal of Stored Products Research, vol. 32, no. 3, pp. 239–246, 1996.
101. S. S. Bedi and M. Singh, "Microwaves for control of stored grain insects," National Academy of Science Letter, vol. 15, no. 6, pp. 195–197, 1992.
102. D. Sanchez-Hernandez, J. V. Balbastre, and J. M. Osca, "Microwave energy as a viable alternative to methyl bromide and other pesticides for rice disinfection of industrial processes," in Proceedings of International Conference on Alternatives to Methyl Bromide, pp. 159–162, Sevilla, Spain, March 2002.
103. W. R. Halverson, T. S. Bigelow, and S. L. Halverson, "Design of High-Power Microwave Applicator for the Control of Insects in Stored Products," in Proceedings of the Annual International Meeting, pp. 27–30, American Society for Agricultural Engineers, Las Vegas, Nev., USA, July 2003, Paper No. 036156.
104. R. Vadivambal, D. S. Jayas, and N. D. G. White, "Wheat disinfestation using microwave energy," Journal of Stored Products Research, vol. 43, no. 4, pp. 508–514, 2007.
105. R. Vadivambal, O. F. Deji, D. S. Jayas, and N. D. G. White, "Disinfestation of stored corn using microwave energy,"

Agriculture and Biology Journal of North America, vol. 1, no. 1, pp. 18–26, 2010.

106. R. Singh, K. K. Singh, and N. Kotwaliwale, "Study on disinfestation of pulses using microwave technique," Journal of Food Science and Technology, vol. 49, no. 4, pp. 505–509, 2012.

107. R. Pande, H. N. Mishra, and M. N. Singh, "Microwave drying for safe storage and improved nutritional quality of green gram seed (Vigna radiata)," Journal of Agricultural and Food Chemistry, vol. 60, pp. 3809–3816, 2012.

108. S. L. Halverson, W. E. Burkholder, T. S. Bigelow, E. V. Norsheim, and M. E. Misenheimer, "High-power microwave radiation as an alternative insect control method for stored products," Journal of Economic Entomology, vol. 89, no. 6, pp. 1638–1648, 1996.

109. H. H. Webber, R. P. Wangner, and A. G. Pearson, "High frequency electric fields as lethal agents for insects," Journal of Economics Entomology, vol. 39, pp. 481–498, 1946.

110. S. O. Nelson, "Dielectric properties of agricultural products: measurements and applications," IEEE Transactions on Electrical Insulation, vol. 26, no. 5, pp. 845–869, 1991.

111. S. O. Nelson, "Radio-frequency and microwave dielectric properties of insects," Journal of Microwave Power and Electromagnetic Energy, vol. 36, no. 1, pp. 47–56, 2001.

112. S. O. Nelson and W. K. Whitney, "Radiofrequency electric fields for stored grain insect control,"Transactions of the American Society of Agricultural Engineers, vol. 3, no. 2, pp. 133–137, 1960.

113. S. O. Nelson, L. E. Stetson, and J. J. Rhine, "Factors Influencing Effectiveness of Radiofrequency Electric Fields for Stored-Grain Insect Control," Transactions of the American Society of Agricultural Engineers, vol. 9, pp. 809–815, 1966.

114. S. O. Nelson, "Electromagnetic Energy," in Pest Control, Biological, Physical and Selected Chemical Methods, W. W. Kilgore and R. L. Doutt, Eds., pp. 89–145, Academic Press, New York, NY, USA, 1967.

115. B. Alfaifi, J. Tang, B. Rasco, S. Sablani, and Y. Jiao, "Radio frequency disinfestation treatments for dried fruit: dielectric

properties," LWT—Food Science and Technology, vol. 50, no. 2, pp. 746–754, 2013.

116. J. N. Ikediala, J. Tang, and T. Wig, "A heating block system for studying thermal death kinetics of insect pests," Transactions of the American Society of Agricultural Engineers, vol. 43, no. 2, pp. 351–358, 2000.
117. V. M. Rashkovan, N. A. Khizhnyak, A. V. Basteev, L. A. Bazyma, L. Niño de Rivera, and I. A. Ponomaryova, "Interaction of electromagnetic waves with granular agricultural product and insects,"Journal of Microwave Power and Electromagnetic Energy, vol. 38, no. 4, pp. 1–12, 2003.
118. J. Tang, "Dielectric Properties of foods," in Microwave Processing of Foods, H. Schubert and M. Regier, Eds., CRC Press, Cambridge UK, 2005.
119. A. C. Metaxas and R. J. Meredith, Industrial Microwave Heating, Peter Peregrinus, 1983.
120. G. J. Hallman and J. L. Sharp, "Radio Frequency Heat Treatments," in Quarantine Treatments For Pests of Food Plants, J. L. Sharp and G. J. Hallman, Eds., pp. 165–170, Westview Press, San Francisco, Calif, USA, 1994.
121. L. E. Campana, M. E. Sempe, and R. R. Filgueira, "Physical, chemical and baking properties of wheat dried with microwave energy," Cereal Chemistry, vol. 70, no. 6, pp. 760–762, 1993.
122. M. V. Bhaskara Reddy, G. S. V. Raghavan, A. C. Kushalappa, and T. C. Paulitz, "Effect of microwave treatment on quality of wheat seeds infected with Fusarium graminearum," Journal of Agricultural Engineering Research, vol. 71, no. 2, pp. 113–117, 1998.
123. J. Kaasova, B. Hubackova, P. Kadlec, J. Prihoda, and Z. O. Bubnik, "Chemical and biochemical changes during microwave treatment of wheat," Czech Journal of Food Science, vol. 20, pp. 74–78, 2002.
124. M. A. K. Hamid and R. J. Boulanger, "A new method for the control of moisture and insect infestations of grain by microwave power," Journal of Microwave Power, vol. 4, no. 1, pp. 11–18, 1969.

125. D. P. Locatelli and S. Traversa, "Microwaves in the Control of Rice Infestants," Italian Journal of Food Science, vol. 1, no. 2, p. 62, 1989.
126. J. L. Sharp and R. G. McGuire, "Control of Caribbean fruit fly (Diptera: Tephritidae) in navel orange by forced hot air," Journal of Economic Entomology, vol. 89, no. 5, pp. 1181–1185, 1996.
127. K. C. Shellie and R. L. Mangan, "Disinfestation: effect of non-chemical treatments on market quality of fruit," in Postharvest Handling of Tropical Fruits, B. R. Champ, Ed., ACIAR Proceedings, pp. 304–310, 1994.
128. J. A. Johnson, K. A. Valero, S. Wang, and J. Tang, "Thermal death kinetics of red flour beetle (Coleoptera: Tenebrionidae)," Journal of Economic Entomology, vol. 97, no. 6, pp. 1868–1873, 2004.
129. K. C. Shellie and R. L. Mangan, "Tolerance of red fleshed grapefruit to a constant or stepped temperature, forced-air quarantine heat treatment," Postharvest Biology and Technology, vol. 7, no. 1-2, pp. 151–159, 1996.
130. D. R. Wilkin and G. Nelson, Control of Insects in Confectionery Walnuts Using Microwaves, vol. 87, British Crop Protection Council Monograph, Berks, UK, 1987.
131. A. Zouba, O. Khoualdia, A. Antoneo Diaferia, V. Valereo Rosito, H. Bouabidi, and B. Chermiti, "Microwave Treatment for Postharvest Control of the Date Moth Ectomyelois ceratoniae," Tunisian Journal of Plant Protection, vol. 4, no. 2, pp. 173–183, 2012.
132. S. Ryynanen, H. Tuorila, and L. Hyvonen, "Perceived temperature effects on microwave-heated meals and meal components," Food Service Technology, vol. 1, pp. 141–148, 2001.
133. D. S. Lee, D. Shin, and K. L. Yam, "Improvement of temperature uniformity in microwave-reheated rice by optimizing heat/cold cycle," Food Service Technology, vol. 2, no. 2, pp. 87–93, 2002.
134. N. Sakai, C. Wang, S. Toba, and M. Watanabe, "An analysis of temperature distributions in microwave heating of foods with non-uniform dielectric properties," Journal of Chemical Engineering of Japan, vol. 37, no. 7, pp. 858–862, 2004.
135. A. Manickavasagan, D. S. Jayas, and N. D. G. White, "Non-uniformity of surface temperatures of grain after microwave

treatment in an industrial microwave dryer," Drying Technology, vol. 24, no. 12, pp. 1559–1567, 2006. V

136. S. Gunasekaran and H. W. Yang, "Effect of experimental parameters on temperature distribution during continuous and pulsed microwave heating," Journal of Food Engineering, vol. 78, no. 4, pp. 1452–1456, 2007.
137. S. S. R. Geedipalli, V. Rakesh, and A. K. Datta, "Modeling the heating uniformity contributed by a rotating turntable in microwave ovens," Journal of Food Engineering, vol. 82, no. 3, pp. 359–368, 2007.
138. A. Manickavasagan, D. S. Jayas, and N. D. G. White, "Germination of wheat grains from uneven microwave heating in an industrial microwave dryer," Canadian Biosystems Engineering, vol. 49, pp. 323–327, 2007.
139. J. M. Hill and M. J. Jennings, "Formulation of model equations for heating by microwave radiation,"Applied Mathematical Modelling, vol. 11, pp. 369–319, 1993.
140. Z. Songping, Z. Yinglong, and T. R. Marchant, "A DRBEM model for microwave heating problems,"Applied Mathematical Modelling, vol. 19, no. 5, pp. 287–297, 1995.
141. A. Ayvaz and S. Karabörklü, "Effect of cold storage and different diets on Ephestia kuehniella Zeller,"Journal of Pest Science, vol. 81, no. 1, pp. 57–62, 2008.
142. S. Gasemzadeh, A. Pourmirza, M. Safaralizadeh, and M. Maroufpoor, "Effect of microwave radiation and cold storage on Tribolium castaneum Herbst (Coleoptera: Tenebrionidae) and Sitophilus oryzae L. (Coleoptera: Curculionidae)," Journal of Plant Protection Research, vol. 50, no. 2, pp. 140–145, 2010.
143. O. Valizadegan, A. A. Pourmirza, and M. H. Safaralizadeh, "The impact of microwaves irradiation and temperature manipulation for control of stored-products insects," African Journal of Biotechnology, vol. 10, no. 61, pp. 13256–13262, 2011.
144. S. M. El-Naggar and A. A. Mikhaiel, "Disinfestation of stored wheat grain and flour using gamma rays and microwave heating," Journal of Stored Products Research, vol. 47, no. 3, pp. 191–196, 2011.

145. M. Amjad and M. Akbar Anjum, "Effect of gamma radiation on onion seed viability, germination potential, seedling growth and morphology," Pakistan Journal of Agriculture, vol. 39, pp. 202–206, 2002.

146. N. C. Doty and C. W. Baker, "Microwave conditioning of hard red spring wheat. I. Effects of wide power range on flour and bread quality," Cereal Chemistry, vol. 54, pp. 717–727, 1977.

147. S. G. Walde, K. Balaswamy, V. Velu, and D. G. Rao, "Microwave drying and grinding characteristics of wheat (Triticum aestivum)," Journal of Food Engineering, vol. 55, no. 3, pp. 271–276, 2002.

148. V. Velu, A. Nagender, P. G. Prabhakara Rao, and D. G. Rao, "Dry milling characteristics of microwave dried maize grains (Zea mays L.)," Journal of Food Engineering, vol. 74, no. 1, pp. 30–36, 2006.

149. R. Vadivambal, D. S. Jayas, and N. D. G. White, "Mortality of stored-grain insects exposed to microwave energy," Transactions of the ASABE, vol. 51, no. 2, pp. 641–647, 2008.

150. S. O. Nelson and L. E. Stetson, "Possibilities for vontrolling insects with microwaves and lower frequency RF energy," IEEE Transactions on Microwave Theory and Techniques, vol. 22, no. 12, pp. 1303–1305, 1974.

Chapter 3

Storage Insects on Yam Chips and Their Traditional Management in Northern Benin

Y. L. Loko[1], A. Dansi[1,2], M. Tamo[3], m A. H. Bokonon-Ganta[4], P. Assogba[2], M. Dansi[1], R. Vodouhè[5], A. Akoegninou[6], and A. Sanni[7]

[1]Laboratory of Agricultural Biodiversity and Tropical Plant breeding (LAAPT), Faculty of Sciences and Technology (FAST-Dassa), University of Abomey-Calavi (UAC), Cotonou, Benin

[2]Crop, Aromatic and Medicinal Plant Biodiversity Research and Development Institute (IRDCAM), 071 BP 28 Cotonou, Benin

[3]Institut International d'Agriculture Tropicale (IITA), 08 BP 0932 Cotonou, Benin

[4]Service de la Protection des Végétaux et du Contrôle Phytosanitaire, Direction de l'Agriculture, BP 58 Porto-Novo, Benin

[5]Bioversity International, Office of West and Central Africa, 08 BP 0932 Cotonou, Benin

[6]National Herbarium, Department of Botany and Plant Biology, Faculty of Sciences and Technology (FAST), University of Abomey-Calavi (UAC), , Cotonou, Benin

[7]Laboratory of Biochemistry and Molecular Biology, Faculty of Sciences and Technology

ABSTRACT

Twenty-five villages of Northern Benin were surveyed to identify the constraints of yam chips production, assess the diversity of storage insects on yam chips, and document farmers' perception of their impacts on the stocks and their traditional management practices. Damages due to storage insects (63.9% of responses) and insufficiency of insect-resistant varieties (16.7% of responses) were the major constraints of yam chips production. Twelve insect pest species were identified among which Dinoderus porcellus Lesne (Coleoptera, Bostrichidae) was by far the most important and the most distributed (97.44% of the samples). Three predators (Teretrius nigrescens Lewis, Xylocoris flavipes Reuter, and Alloeocranum biannulipes Montrouzier & Signoret) and one parasitoid (Dinarmus basalis Rondani) all Coleoptera, Bostrichidae were also identified. The most important traditional practices used to control or prevent insect attack in yam chips were documented and the producers' preference criteria for yam cultivars used to produce chips were identified and prioritized. To further promote the production of yam chips, diversification of insect-resistant yam varieties, conception, and use of health-protective natural insecticides and popularization of modern storage structures were proposed.

INTRODUCTION

In economic terms, yams (Dioscorea spp.) are the world's fourth most important tuber crop after potatoes, cassava, and sweet potatoes [1]. They are cultivated in most tropical countries but especially in West Africa, where over 95% of the world's output is produced [1, 2]. They are the main source of carbohydrate for millions of people. In West Africa, many yam species are cultivated but the African domesticates known as the Guinea yams (D. cayenensis Lam.-D. rotundata Poir. complex) are, however, the most important, most preferred, and widely planted [3].

In Benin, the fourth yam-producing country behind Nigeria, Ivory Coast, and Ghana, yam is among the most important food crops [1] and has economic and sociocultural importance [4]. Yam is seasonal and the fresh tubers are highly perishable. Postharvest losses are very high, ranging from 30% to 85% of the total production [4]. In order to overcome this high perishability of the tubers and the irregularity of its availability throughout the year, yams are traditionally processed into dried chips or cossettes [5–7], hence reinforcing food security [7]. Unfortunately, yam chips are often severely attacked by insects, which sometimes reduce whole yam stocks into powder in very few months [8, 9]. Very little research attention has been given to storage insects attack on yam chips and traditional management practices in Benin. Gnonlonfin et al. [10] reported the existence of many species of insects but focused their study mainly on their population's dynamics in stored yams chips. Consequently, the diversity of the insect species in the yam chips producing zone is still unknown, and farmers' perception of the importance of insect damages in the stocks has never been assessed. Traditional management practices (including the storage structures) used to prevent or control insect infestations have also not been documented. Yam chips are produced from tubers of single-harvest varieties, locally known as "Kokoro" characterized by their numerous small-sized tubers. Within Kokoro yams, many varieties of different agronomic and technological characteristics exist [11] but the yam chips producers' variety preference or selection criteria have never been studied. Knowledge of farmers' selection criteria will be useful in designing concrete breeding programmes that could facilitate the adoption of improved varieties [12].

We report in this paper a survey conducted in the most important yam chips producing zone of Benin in order to

- identify and prioritise the constraints related to the production of yam chips in Benin and farmers' propositions for overcoming these constraints,
- document farmers' perceptions about insect pests on stored yam chips and traditional management practices,
- assess the diversity and the importance of the storage insects species in the most important yam chips production zone of Benin,

- Identify and prioritize the producers' variety preference or selection criteria across study zones and ethnic groups for popularization and breeding purposes.

MATERIAL AND METHODS

Study Area

The study was conducted in five districts (Djougou, Copargo, Ouaké, Bassila, and Tchaourou) of the Departments of Donga and Borgou in northern Benin. These districts are known to be the major yam chips production zones of Benin [11, 13]. The inhabitants are members of seven ethnic groups (Figure 1; Table 1) (Ani, Bariba, Lokpa, Nago, Peulh, Taneka, and Yom) and have a very long tradition in processing Kokoro yam tubers into chips or cossettes (Figure 2). The departments of Donga and Borgou are located in a semiarid agroecological zone characterized by unpredictable and irregular rainfall (800–1300 mm/year) with only one rainy season (May to October) and a dry season lasting for more than 5 months sometimes [14]. Mean annual temperatures range from 26°C to 28°C and may exceptionally reach 35°C [15, 16]. Yam production in this area is intensive and essentially based on Kokoro yams [13], which have very variable yields from one season to another due to climatic hazards [13].

Table 1: Administrative localisation of the ethnic areas and sites surveyed

N°	Ethnic areas	Districts	Number of sites	Selected villages
1	Yom	Djougou	6	Déwa, Alfapara, Pélébina, Mone, Gangamou, Dangoussar
2	Lokpa	Djougou	4	Ouarlgou, Yarakéou, Pohomto, Niagba-kabia
3	Ani	Bassila	2	Penessoulou, Saramanga

4	Nago	Bassila	7	Modogui, Ouanou, Papané, Agramarou, Koko, Agbassa, Wari-Maro
5	Taneka	Copargo	3	Kataban, Setrah, Foungou
6	Bariba	Tchaourou	2	Woria, Badékparou
7	Peulh	Tchaourou	1	Gakpenou

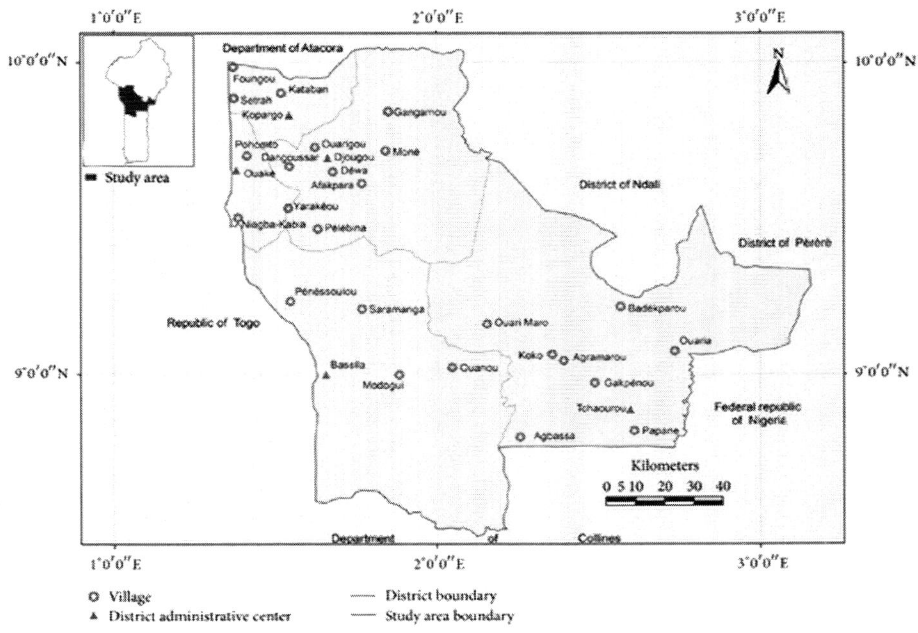

Figure 1: Map of Benin showing the geographical position of the surveyed villages.

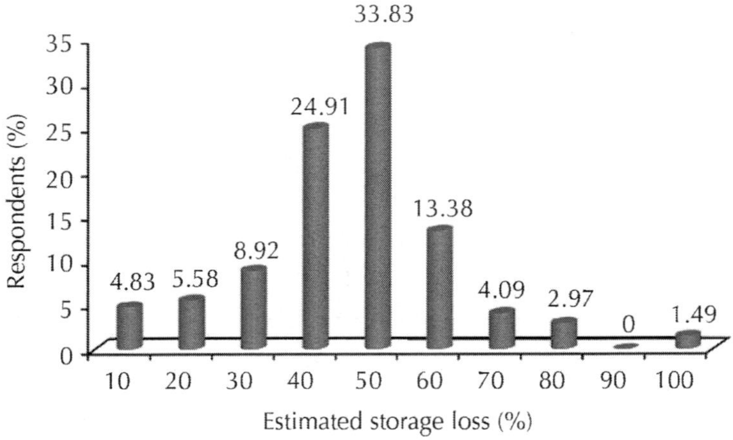

(a)

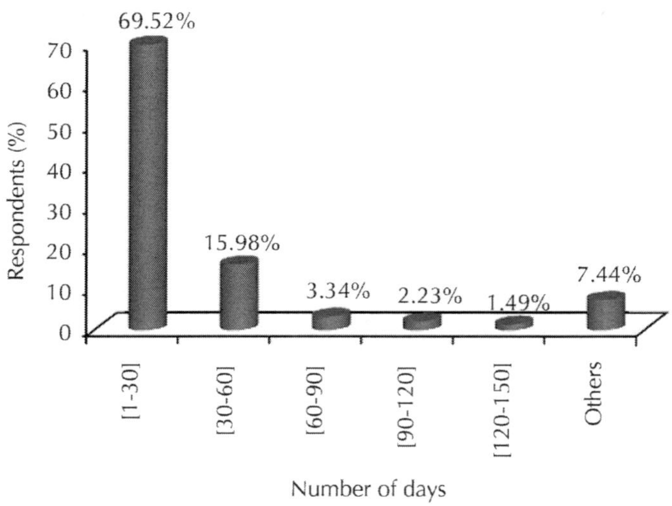

(b)

Figure 2: Farmers' perception of (a) storage loss due to stored yam chips insect pests, (b) the period of infestation of the yam chips.

Site Selection and Survey

Twenty-five villages (Table 1; Figure 1) were randomly selected throughout the study area and its ethnic zones for the survey. Data were collected from the different sites during expeditions through the application of Participatory Research Appraisal tools and techniques such as granary visits, direct observation, focus group discussions, and individual interviews using a questionnaire and the help of translators from each area following Dansi et al. [12]. In each site, local farmers' associations were involved in the study to facilitate the identification of the households to survey and the data collection. Within villages, 10 to 12 households were randomly selected for individual interviews using the transect method described by Dansi et al. [17]. In each household, the interviewee (head of household or his wife or one of his wives in case of polygamy) was selected by mutual agreement with the hosting couple according to Christinck et al. [18]. Apart from the socioeconomic data such as age, gender, and educational level of the interviewees, data collected included the farmers' perceptions of the constraints of yam chips production, the cossette storage structures and practices, the importance of damages caused by insects, the time of the infestation, the farmers' knowledge of the insect species, the traditional management practices on the infested stocks, and the farmers' preference criteria of kokoro varieties used in the production of the chips. Preference criteria of kokoro varieties were identified and prioritized using the matrix scoring technique described by Defoer et al. [19], Adoukonou-Sagbadja et al. [20], and Dansi et al. [12]. In each village, samples (500 g) of infested yam chips were collected directly from two to three randomly selected yam chips storage structures following Mendesil et al. [21] and Koradaa et al. [22]. Initial weights of the samples to be collected were taken using the numerical balance (model SF-400). Infested samples collected were preserved in plastic containers with perforated lids to allow for ventilation. With the aid of a plant taxonomist at the national herbarium of the University of Abomey-Calavi, insecticides and/or insect repulsive plants used in processing reported by interviewees we sampled and their scientific names were determined using the analytical flora of Benin [16].

Incubation of the Samples and Isolation and Identification of the Insects

The labeled plastic containers containing the samples of the infested chips were incubated for three months under laboratory conditions at temperature of 25–27°C and 70%–80% relative humidity, following the method described by Eze et al. [23]. After the incubation period, the samples were broken into particles of less than 0.5 cm using a hand mortar and the insects were recovered through a 0.25 mm nylon net sieve [24]. Recovered insects were counted and conserved in a flask containing 70% alcohol for safeguarding and identification. Species' identification was done at the Biodiversity Center of the International Institute of Tropical Agriculture (IITA-Benin).

Statistical Analysis

Data were analyzed through descriptive statistics (frequencies, percentages, means, etc.) to generate summaries and tables at different (villages, individuals) levels using SAS software [25].

RESULTS

Characteristics of the Respondents

The respondents were in majority (98%) women. Sixty-three (63%) are illiterate and 47% attended primary school only. Their ages varied from 17 to 60 years with an average of 37 years. In all the households surveyed, yam chips were produced for either home consumption and for the market (95.2% of the respondents) or for home consumption only (4.8%).

Constraints of Yam Chips Production

Six constraints (Table 2) related to yam chips production in Benin were recorded. They were all directly or indirectly linked to the storage of the chips. Among them, damages due to storage insects were the most important (63.9% of responses), followed by insufficiency of insect-resistant varieties (16.7% of responses) and the lack of natural human health preserving insecticides (10.2% of responses). The other three constraints (lack of organised markets, low availability of fresh kokoro yam tubers, and the lack of appropriate and specific storage structures) were of very low importance (only 1.1% to 4.5% of responses). The majority of the yam chips producers (72.12%) estimated at 40%–60% the importance of the damage caused by the storage insects on the yam ships (Figure 2(a)). This however depends on the variety used, the conservation structure and the drying level of the chips. For the great majority (92.94%) of the respondents, the infestation of the cossettes in stock occurred during the first two months (Figure 2(b)). In order to minimize these constraints, yam chips producers proposed six key solutions (Table 3) including diversification of good storage insect-resistant kokoro yam (30.2% of responses), development of a natural human health preserving insecticides (24.2% of responses), enhancement of the production of kokoro yam (21.2% of responses), and development of fast drying areas for the yam chips (19.5% of responses).

Table 2: Yam chips production constraints in Benin

Constraints	Percentage of responses
Damages caused by storage insects	63.9
Insufficiency of insect-resistant varieties	16.7
Lack of natural human health preserving insecticides	10.2
Lack of appropriate and specific storage structures	4.5
Insufficient availability of fresh kokoro yam tubers	3.6
Lack of organised markets	1.1

Table 3: Solutions for the constraints and their importance as proposed by the interviewees in the study area

Solutions	Importance (% of responses)
Diversification of good storage insect-resistant kokoro yam	30.2
Development of a natural human health preserving insecticides	24.2
Enhancement of the production of kokoro yam	21.2
Development of fast drying areas for the yam chips	19.5
Development of efficient and specific yam chips storage structures	3.3
Establishment of a well-organised yam chips good market	1.6

Farmers' Knowledge of the Insect Pests Damaging Yam Chips in Stocks and Diversity Assessment

In the study area, all the storage insects were traditionally classified in a single group named Benonkpé in Ani,Kokolibo in Nago, Doridji in Peulh, Gbénénoukokonou in Bariba, Dresse in Yom, and Poucasse in Lokpa and Taneka. All these six vernacular names literally mean beetles. Farmers reported that these beetles act by penetrating the chips and drastically reducing their internal parts into powdery waste (Figure 3). Although interviewees recognized storage insects as major constraint in yam chips production, only 47% of them were able to differentiate some species. The few respondents who attempted to identify yam chips insect pests based their identification mainly on the colour (45.12% of responses) and the relative size (32.18% of responses) of the insects and on the symptoms of the damage they caused (22.7% of responses).

Figure 3: Yam chips with insect infestation in Benin.

The diversity analysis conducted on the total of 78 samples collected and incubated revealed 12 species of insects belonging to four orders (Table 3) which are Coleoptera (eight species), Hemiptera (two species), Hymenoptera (one species), and Psocoptera (undetermined species). Species of the order of Coleoptera were the most numerous and the most represented in the samples. On average 223 insects of the order Coleoptera were counted by 500 g of yam chips against 11.4 for all the other orders put together (Table 4). Among the species identified Dinoderus porcellus Lesne was the most represented. It was found in 97.44% of the collected samples and was also the most abundant in all the samples in which it was found (Table 4). This was followed by the species Tribolium castaneum (Herbst), detected in 52.56% of the samples (Table 4). The other species were found in only 2 to 10 samples out of the 78 samples collected and in very few numbers. Among the 12 species of insects identified, three (Teretrius nigrescens Lewis, Xylocoris flavipes Reuter, and Alloeocranum biannulipes Montrouzier & Signoret) were predators and one (Dinarmus basalis Rondani) was a parasitoid.Xylocoris flavipes was found in 12.82% of the samples and appeared to be the most abundant predator.

Table 4: Results of the samples incubated at the laboratory showing the species of insects used and their relative abundance

Types of insects	Infested samples (%)	Average count for 500 g	Percentage of abundance (500 g)	Rank
Dinoderus porcellus	76 (97.44)	208.72	89.03	1
Psocopteraspp.	19 (24.36)	8.83	3.77	2
Tribolium castaneum	41 (52.56)	6.12	2.61	3
Lasioderma serricorne	4 (5.13)	3.3	1.41	4
Sitophilus zeamais	10 (12.82)	2.57	1.09	5
Xylocoris flavipes	10 (12.82)	2.46	1.05	6
Cryptolestes pusillus	10 (12.82)	1.26	0.54	7
Carpophilus dimidiatus	7 (8.97)	0.85	0.36	8
Teretrius nigrescens	5 (6.41)	0.13	0.05	9
Carpophilus binotatus	3 (3.85)	0.09	0.04	10
Alloeocranum biannulipes	5 (6.41)	0.09	0.04	11
Dinarmus basalis	2 (2.56)	0.03	0.01	12

Traditional Yam Chips Storage Systems and Duration of the Conservation

In all the households surveyed, yam chips were stored inside houses and rooms in various containers. The great majority (97.77%) of producers used maize bags (made with synthetic materials) of various sizes as containers, depending on the quantity of chips to be conserved. Only few producers (2.23%) preferred to store in large-sized and hermetically closed plastic buckets, jars, or barrels to prevent insect infestations. No specific structure was dedicated to storage of yam chips. Interviewees reported that storage period varied from 1 to 13 months with an average of 8 months. For 49.82%, 31.97%, and 18.21% of the respondents, storage duration of yam chips varied between 1 and 5 months, 5 and 10 months, and 10 and 15 months, respectively.

Traditional Control Systems of Yam Chips Storage Insects Pests

Under traditional storage conditions, interviewees used seven strategies to reduce losses due to insects (Table5). Among those, the most important were regular inspection and exposure of chips to sunlight to repel insects (35.93% of responses), use of insect-resistant varieties (26.80% of responses), and use of insecticide and/or insect's repulsive plants during preparation (26.45% of responses). The other strategies such as shaking of the yam chips to remove insects along powdery waste, use of insecticides, treatment with pepper powder, and minimising frequent opening of storage structures to avoid entrance of the insects were poorly used (Table 5). According to interviewees, severely infested stocks of yam chips were sold (66.17% of responses), used for home consumption only (24.54% of responses), or simply thrown away (9.29% of responses).

Table 5: Farmers' management practices for the control of yam chips insect pests

Management practices	Importance (% of responses)
Exposure of the infested yam chips to sun	35.93
Use of insect-resistant varieties	26.80
Use of insecticide and/or insect's repulsive plants during preparation	26.45
Sifting of yam chips to remove insects along powdery waste	8.59
Use of insecticides	1.12
Treatment with pepper powder	0.74
Minimising frequent opening of storage structures to avoid entrance of the insects	0.37

The study revealed that eight species of plant were used to prevent infestation of the yam chips or to control insect pests (Table 6). Among these species, three (Bridelia ferruginea, Blighia sapida, and Khaya senegalensis) were reported to be insecticide while four (Piliostigma thonningii, Lophira lanceolata, Tectona grandis, andSorghum bicolor) were said to be dyes and insect repulsive (Table 6). Cassava leaves were also used during the parboiling to harden the chips. Piliostigma thonningii and Sorghum bicolor were known and used across all the ethnic groups while the other species apart from Tectona grandis were each used in only one ethnic group. The number and types of species of plants varied from one ethnic group to another. Yom and Peulh used only two species of plant, Lokpa and Ani used four species, and Nago used five. For the different plant species identified, the plant parts (leaves or bark) used, the application or treatment methods (infusion or fumigation), and the frequencies of utilization across ethnic groups are summarized in Table 6.

Table 6: List of plants used to protect yam chips against storage insect pests and their utilisation methods

Species	Part used	Role	Method of application	Percentage of farmers using the plants across ethnic groups						
				Peulh	Nago	Ani	Taneka	Bariba	Lokpa	Yom
Piliostigma thonningii	Leaf/bark	Dye	Infusion	42.86	19.40	15	30.30	16.67	26.47	30.30
Lophira lanceolata	Leaf	Dye	Infusion	—	7.46	—	—	—	—	—
Blighia sapida	Leaf	Insecticide	Infusion	—	5.97	—	—	—	—	—
Bridelia ferruginea	Leaf/bark	Insecticide	Infusion	—	—	5	—	—	—	—
Khaya senegalensis	Bark	Insecticide	Fumigate	—	—	—	—	11.11	—	—
Tectona grandis	Leaf	Dye	Infusion	—	19.40	5	12.12	—	14.71	—
Manihot esculentus	Leaf	Hardening of the yam chips	Infusion	—	—	—	—	—	2.94	—
Sorghum bicolor	Stem/oil cakes	Dye	Infusion	57.14	47.77	75	57.58	72.22	55.88	69.70

Throughout the study zone, 37 kokoro yam cultivars used to produce chips were listed as tolerant to storage insects. The number of cultivars reported varies across ethnic areas. Eight cultivars were reported with the Nago, Bariba, and Lokpa, seven with the Yom, four with the Taneka, and only two with the Ani (Table 7). In each ethnic area, certain cultivars were more common. With the Nago ethnic group, Oguidigbo, Adakada, Tabané, and Omonya were the most important while with the Bariba ethnic group, the most listed cultivars were Otoukpannan, Tchakatchaka, and Yakanougo (Table 6).

Table 7: Kokoro yam cultivars tolerant to storage insect pests and their importance across ethnic areas

Ethnic areas	Insect-resistant varieties	Importance (number of farmers)
Ani	Demkpenai	14
	Awanawou	11
	Otoukpannan	18
	Tchakatchaka	16
	Yakanougo	15
Bariba	Omonya	8
	Singor	6
	Ankakorouwoura	1
	Gaboubaba	1
	Kourakourogouroko	1
	Azowi	17
	Iootchra	14
	Moghoun	12
Lokpa	Kounto	10
	Kparokoumè	8
	Soprova	6
	Tougbana	4

	Tédoman	1
	Oguidigbo	54
	Adakada	45
	Tabané	41
Nago	Hounbonon	33
	Kokorogbambe	26
	Kokorolakolako	18
	Kokoroagbalè	13
	Adjawoungbo	5
	Atawouraï	27
Taneka	Souwoukou	19
	Gréé	7
	Djèssoumè	4
	Koutonouman	51
	Biboï	44
Yom	Assinakpeina	39
	Adjôgba	17
	Mouhame	3
	Ayè	2
	Satchila	2

Farmer's Preferences Criteria for kokoro Yam Cultivars for Chips Production

Throughout the study zone, kokoro yam cultivars, used for chips production, were selected among the existing diversity based on eight criteria. Among them, the quality of the paste made with the flour, the storage aptitude of the chips, and the quality of the Wassa-Wassa (local couscous made with yam chips' flour) were the most important and represent altogether 74.08% of the responses (Table 8). The number and importance of the criteria also varied across ethnic groups. With the Nago and the Peulh, storage aptitude was the most important criterion while with the other ethnic groups the quality of the paste came at the first position (Table8). The quality of Wassa-Wassa, which was the third most important criterion among the Lokpa, Ani and Yom people, was not even mentioned by the Bariba people. Similarly, the fast drying quality of the cultivars, which was important to the Nago, the Taneka, and the Ani, was not listed among the Peulh people and had very low values with the other ethnic groups. While all the eight criteria were listed by the Lokpa ethnic group, all but one was recorded with the Taneka and the Yom and only four were identified with the Peulh (Table 8).

Table 8: Famer's preference criteria of good kokoro yam cultivar for chips production in the study area and across ethnic groups

Preferences criteria	Study area(% of responses)	Ethnic groups (% of responses)						
		Nago	Peulh	Bariba	Taneka	Lokpa	Yom	Ani
Quality of the paste	35.62	23.03	21.43	41.38	38.15	31.25	35.65	32.14
Storage aptitude of the chips	26.5	31.99	57.14	37.93	24.5	23.96	22.61	28.57
Quality of Wassa-Wassa	11.96	4.11	7.14	0	3.54	22.92	13.04	17.86
Colour of the paste	9.67	12.33	14.29	13.79	2.7	6.25	13.04	3.57
Flour richness of the yam chips	3.05	2.74	—	—	7.9	7.29	0	—
Crushing facility of the chips	1.02	—	—	—	2.6	2.08	0.87	—
Taste of the paste	4.83	—	—	3.45	—	5.21	10.44	3.57
Fast drying	7.35	25.8	—	3.45	20.61	1.04	4.35	14.29

DISCUSSION

The respondents were in majority women. This can be explained by the fact that in all ethnic groups of the study area, women were the sole processors of kokoro yam tubers into chips. The few males interviewed responded on behalf of their wives, who gave way to them out of respect. The culture of the people was also evident in the yam production system where tasks had been traditionally divided according to gender. Men were in charge of the most important activities in terms of labour requirements, while foods processing and transformation of the yam tubers into chips, among other activities, were devoted to women [4]. According to Bricas and Vernier [26], the commercialization of yam chips is by far more economically profitable than the one of fresh tubers. This could justify the importance that women in the study area gave to commercialization as a means of substantially improving their household income.

Among the constraints related to yam chips production in the study area, damages caused by storage insects stood out as the most important. Similar results were reported by Osuji [27] and Adedire and Gbaye [28] in Nigeria. The importance of the damages raised by the respondents is key indicator of the urgent necessity to develop control strategies against the storage insects.

Sun drying of infested chips was the major method used by farmers to control these insect pests. This method, which is the oldest known technique of conservation of the agro-alimentary products, also presents several disadvantages. Chalal et al. [29] reported that sun drying directly exposes the products to dust and to ultraviolet rays which can cause the deterioration of food vitamins. Among the solutions proposed by farmers were diversification of good storage insect-resistant kokoro yams and development of natural human health preserving insecticides. These two propositions, which call the attention of plant geneticist and breeders on one hand and industrial chemists on the other, indicate that producers are very concerned about their health. The numerous cases of food poisoning that were associated with the use of cotton insecticides on yam chips recorded these last years in the country and which led to the death of many persons may have contributed to this health consciousness. In Nigeria, Adedoyin et al. [30] and Adeleke [31] also reported poisoning due to the consumption of yam flour (treated with insecticide) in some families in Ilorin and Kano.

Our study revealed that in the different samples of infested yam chips collected and analyzed, Dinoderus porcellus was the most represented. This species which is known to be mostly associated with dried yams [32] has already been reported as the most important pest of stored yam chips in Nigeria [27, 28, 32]. Dinoderus porcellus particularly infests well-dried chips [27, 28]. Therefore, it is possible that the few samples, in which it was absent, were not well dried or had relatively higher moisture contents. The presence of Psocoptera spp.,Carpophilus dimidiatus, and Carpophilus Binotatus in the samples without Dinoderus porcellus supports this hypothesis as they are known as insects associated with wet food products [33]. Tribolium castaneum andPsocoptera spp. were also found in not negligible number of the samples. The red flour beetle, Tribolium castaneum, is a common and one of the most important stored product pests associated with a wide range of durable commodities (barley, bran, cacao, ginger, maize, millet, cassava chips, nutmeg, peanut, pepper, rice, sorghum, etc.) and food-processing facilities worldwide [34–36] were also found. Its presence in the samples examined is not surprising as it has already been reported by Vernier et al. [9], Soldati et al. [37], and Oni and Omoniyi [32]. In some yam chips samples collected outside our study area, Vernier et al. [9, 38] and Gnonlonfin et al. [10] identified five other species which were not found in our studies. These includedDinoderus bifoveolatus (Wollaston), Palorus subdepressus (Wollaston), Rhyzopertha dominica (F.), C. quadricollis (Guérin-Méneville), Gnatocerus maxillosus (F.), and Prostephanus truncatus. In order to have an exhaustive list of all the stored-products insect pests associated with yam chips in Benin and map their geographical distribution, further studies need to be conducted by including the remaining part of the country. A good knowledge of the diversity of the species will be of great utility for the yam breeders who may like to select kokoro cultivars producing tubers that are tolerant to storage insect pests. For example, The Laboratory of Agricultural Biodiversity and Tropical Plant Breeding of the University of Abomey-Calavi (Benin) and the Global Crop Diversity Trust (Rome, Italy) are currently introducing yam chips technology to the arid zone of the department of Atakora (far northwest of Benin), where the environment is quite suitable for fast drying of the chips. To succeed, however, it will be necessary to reckon with kokoro cultivars tolerant to storage insect pests.

To control insect pests and diseases in crops, Integrated Pest Management (IPM) is recommended [39]. IPM promotes biological control based on the use of the natural enemies of pests (predators and parasitoids) and the genetic control through growing of pest and disease tolerant or resistant cultivars [40]. Among the natural enemies encountered in the infested yam chips, Xylocoris flavipes is known as an effective polyphagous predator of eggs, larvae and chrysalis of coleopteran insects [41]. This natural enemy is also a predator of larvae of T. castaneum [42] and is frequently associated with the insects of cereals stocks [43]. According to Helbig [44], X. flavipes has some interesting biological characteristics that make it a potential control agent of storage insect pests. Unfortunately, it was reported that X. flavipes only eliminates populations of small-sized insects, but not larger insects or insects with internal feeding such as D. porcellus [45]. A. biannulipes on the other hand (Montrouzier & Signoret) is known as a predator of the large-sized storage insect pests including L. serricorne (F.) and Tribolium castaneum [46–48]. A biological control program, combining these predators, will be useful in eliminating various types of insects and will help control the insect pests' complex associated with yam chips.

It appeared from our study that yam chips producers also used diverse plants to protect chips against insect attacks. Phytochemical studies conducted by Dumaine et al. [49] revealed that none of the four plant species (L. lanceolata, T. grandis, P. thonningii, and B. ferruginea) used in the study areas has insecticidal or insect repulsing effects. However, Akinpelu and Obuotor [50] found that P. thonningii bark extract has a bactericidal activity which is also important for improving the sanitary conditions of the chips. Among the plants used,Blighia sapida, Bridelia ferruginea, and Khaya senegalensis are even believed, and rightly so, to have insecticide properties by the Nago, the Ani, and the Bariba people (Table 6). In fact, the bark extract of K. senegalensis has been proved to be antifungal [51], antibacterial [52, 53], and insect antifeedant [54]. Mitchell and Ahmad [55] reported that B. sapida has acaricide and insecticide properties. Similarly it has been shown that all the fruit components (skin, aril and granulates, oil) of this plant have repulsive properties against stock insects such asCallosobruchus maculatus, Cryptolestes ferrugineus, T. castaneum, and S. zeamais [55–58]. Experiments should be conducted to assess the effects of the extract of these three species on insect pests that damage stored yam chips.

Chips producers reported that the importance of the damages is a function of the yam cultivars used and listed 37 kokoro yam landraces producing tubers rarely attacked by the storage insect pests. Due to the existence of numerous synonymies in farmer-named yam cultivars [11], these listed landraces may not all correspond to 37 different genotypes. Therefore, agromorphological characterization coupled with molecular analysis should be carried out to identify duplicates and establish the equivalence between recorded names following Tamiru et al. [59] and Kombo et al. [60]. Moreover, and as recommended by Vernier et al. [38], it will be also important to assess by a well-elaborated trial the effectiveness of the tolerance of the chips derived from the tubers of these varieties to storage insect pests. The use of resistant varieties remains the most economically profitable and the best healthy method of combating chips storage insect pests. Because of this, kokoro yams in the chips production zone should be strengthened with more high yielding cultivars that are suitable for chips and resistant to storage insect pests. According to Dansi et al. [11], such cultivars exist in the traditional agriculture and could be identified through participatory evaluation. Within the existing diversity, cultivars to be used for the chips are selected based on diverse criteria, among which those related to the quality of the foods (Wassa-Wassa; paste) made with the yam chips flour and the technological characteristics of the chips are the most important (66.66% of the responses). This result is expected because in Benin, chips are only made and used for food purposes. In the preference criteria identified, aspects related to conservation come in second position indicating that producers really give particular importance to insect damages. The variation of the preference criteria noted across ethnic groups is frequent and has been already reported in many crops such as cowpea [61], banana and plantain [62], maize [63], telf [64], sorghum [65], yam [11], and even fonio [12]. The fast drying criteria importantly raised by the Nago, the Taneka, and the Ani ethnic groups should be seriously considered as it influences the hygienic quality of the chips and their market value. Nago, Taneka, and Ani people mostly produced chips for economic purposes through commercialization. One understands therefore how important fast drying could be to them.

CONCLUSIONS

This study has allowed us to identify several constraints that hamper yam chips production in northern Benin. Attacks by storage insects were the major constraints identified. Yam chips were infested by various insects, of which the most important was Dinoderus porcellus. Several plants are traditionally used to fight these insects. Following farmers' requests, efforts should be directed towards diversification of good kokoro cultivars which are tolerant to storage insects. In this framework and to identify such cultivars, we recommend the participatory evaluation of existing kokoro yam, the identification of duplicates, and clarification of synonymies and the assessment of the tolerance of the chips manufactured with tubers produced by the identified varieties.

ACKNOWLEDGMENTS

This research was sponsored by the Ministry of Higher Education and Scientific Research of Benin. The authors are grateful to Dr. Goergen Georg, who is in charge of the Insect Museum of the International Institute of Tropical Agriculture (Cotonou, Benin) for his assistance during the identification of the insects and to Dr. H. Yédomonhan (National Herbarium, Department of Botany and Plant biology, University of Abomey-Calavi) for the identification of the plant species. They thank all the chips manufacturers and agricultural technicians they met for fruitful discussions during the survey.

REFERENCES

1. FAO, FAOSTAT Database, Food and Agriculture Organization, Roma, Italy, 2010, http://www.fao.org/.

2. J. M. Babajide, O. Q. Bello, and S. O. Babajide, "Quantitative determination of active substances (preservatives) in Piliostigma thonningii and Khaya ivorensis leaves and subsequent transfer in dry-yam," African Journal of Food Science, vol. 4, no. 6, pp. 382–388, 2010.
3. H. D. Mignouna and A. Dansi, "Yam (Dioscorea ssp.) domestication by the Nago and Fon ethnic groups in Benin," Genetic Resources and Crop Evolution, vol. 50, no. 5, pp. 519–528, 2003.
4. M. N. Baco, S. Tostain, R. L. Mongbo, O. Dainou, and C. Agbangla, "Gestion dynamique de la diversité variétale des ignames cultivées (Dioscorea cayenensis—D. rotundata) dans la commune de Sinendé au nord Bénin," Plant Genetic Resources Newsletter, vol. 139, pp. 18–24, 2004.
5. N. Akissoé, D. J. Hounhouigan, N. Bricas, P. Vernier, C. M. Nago, and O. A. Olorunda, "Physical, chemical and sensory evaluation of dried yam (Dioscorea rotundata) tubers, flour and amala, a flour-derived product," Tropical Science, vol. 41, no. 3, pp. 151–155, 2001.
6. D. J. Hounhouigan, A. P. Kayode, N. Bricas, and C. M. Nago, "Desirable culinary and sensory characteristics of yams in urban Benin," Benin Journal of Agricultural Sciences, vol. 21, no. 12, pp. 2815–2820, 2003.
7. J. M. Babajide, O. O. Atanda, T. A. Ibrahim, H. O. Majolagbe, and S. A. Akinbayode, "Quantitative effect of 'abafe' (Piliostigma thionnigii) and 'agehu' (Khaya ivorensis) leaves on the microbial load of dry-yam 'gbodo'," African Journal of Microbiology Research, vol. 2, pp. 292–298, 2008.
8. E. Ategbo, N. Bricas, J. Hounhouigan et al., "Le développement de la filière chips d'igname pour l'approvisionnement des villes au Nigeria, au Bénin et au Togo," in Actes du Séminaire International Cirad-Inra-Orstom-Coraf, pp. 339–341, Montpellier, France, Juin 1997.
9. P. Vernier, G. Goergen, R. Dossou, P. Letourmy, and J. Chaume, "Utilization of biological insecticides for the protection of stored yam chips," Outlook on Agriculture, vol. 34, no. 3, pp. 173–179, 2005.
10. G. J. B. Gnonlonfin, K. Hell, A. B. Siame, and P. Fandohan, "Infestation and population dynamics of insects on stored cassava

and yams chips in Benin, West Africa," Journal of Economic Entomology, vol. 101, no. 6, pp. 1967–1973, 2008.

11. A. Dansi, H. D. Mignouna, J. Zoundjihékpon, A. Sangare, R. Asiedu, and F. M. Quin, "Morphological diversity, cultivar groups and possible descent in the cultivated yams (Dioscorea cayenensis—D. rotundata) complex in Benin Republic," Genetic Resources and Crop Evolution, vol. 46, no. 4, pp. 371–388, 1999.
12. A. Dansi, H. Adoukonou-Sagbadja, and R. Vodouhè, "Diversity, conservation and related wild species of Fonio millet (Digitaria spp.) in the Northwest of Benin," Genetic Resources and Crop Evolution, vol. 57, no. 6, pp. 827–839, 2010.
13. Y. L. Loko, A. Dansi, C. Linsoussi et al., "Current status and spatial analysis of Guinea yam (Dioscorea cayenensis Lam. -D. rotundata Poir. complex) diversity in Benin," Genetic Resources and Crop Evolution. In press.
14. S. Adam and M. Boko, Le Bénin, Les éditions du Flamboyant / EDICEF, 1993.
15. A. C. Adomou, Vegetation patterns and environmental gradients in Benin: implications for biogeography and conservation [Ph.D. thesis], University of Wageningen, Wageningen, The Netherlands, 2005.
16. A. Akoegninou, W. J. van der Burg, and L. J. G. van der Maesen, Flore Analytique du Bénin, Backhuys Publishers, 2006.
17. A. Dansi, A. Adjatin, H. Adoukonou-Sagbadja et al., "Traditional leafy vegetables and their use in the Benin Republic," Genetic Resources and Crop Evolution, vol. 55, no. 8, pp. 1239–1256, 2008.
18. K. V. Christinck, B. Kshirsagar, E. Weltzien, and P. J. Bramel-Cox, "Participatory methods for collecting germplasm : experiences with famers of Rajasthan, India," Plant Ressources Newsletter, vol. 121, p. 129, 2000.
19. T. Defoer, A. Kamara, and H. De Groote, "Gender and variety selection: farmers› assessment of local maize varieties in Southern Mali," African Crop Science Journal, vol. 5, no. 1, pp. 65–76, 1997.
20. H. Adoukonou-Sagbadja, A. Dansi, R. Vodouhè, and K. Akpagana, "Indigenous knowledge and traditional conservation of fonio

millet (Digitaria exilis, Digitaria iburua) in Togo," Biodiversity & Conservation, vol. 15, no. 8, pp. 2379–2395, 2006.

21. E. Mendesil, C. Abdeta, A. Tesfaye, Z. Shumeta, and H. Jifar, "Farmers› perceptions and management practices of insect pests on stored sorghum in Southwestern Ethiopia," Crop Protection, vol. 26, no. 12, pp. 1817–1825, 2007.
22. R. R. Koradaa, S. K. Naskara, and S. Edison, "Insect pests and their management in yam production and storage: a world review," International Journal of Pest Management, vol. 56, no. 4, pp. 337–349, 2010.
23. S. C. Eze, J. E. Asiegbu, B. N. Mbah, G. C. Orkwor, and R. Asiedu, "Effects of four agrobotanical extracts and three types of bags on the control of insect pests and moulds of stored yam chips," Journal of Agriculture, Food, Environment and Extension, vol. 5, no. 1, pp. 8–12, 2006.
24. A. Delobel, "Les cossettes de manioc, un important réservoir d'insectes des denrées stockées en Afrique centrale," Revue de Zoologie Africaine, vol. 106, no. 1, pp. 17–25, 1992.
25. SAS, "User›s guide," SAS Institute, Cary, NC, USA, 1996.
26. N. Bricas and P. Vernier, Dossier: la transformation de l›igname. Bulletin du réseau TPA N° 18, 5–12, 2000, http://infotpa.gret.org/fileadmin/bulletin/bulletin18.pdf.
27. F. N. C. Osuji, "Observations on beetles attacking dried yams and yam flour from three Nigerian markets," Tropical Stored Products Information, vol. 39, pp. 35–38, 1980.
28. C. O. Adedire and O. A. Gbaye, "Seasonal prevalence and life history of the yam beetle, Dinoderus porcellus (Lesne) (Coleoptera: Bostrichidae)," Nigerian Journal of Experimental and Applied Biology, vol. 3, pp. 323–329, 2002.
29. N. Chalal, A. Bellhamri, and L. Bennamoun, "Étude d'un séchoir solaire fonctionnant en mode direct et indirect," Revue des Energies Renouvelables, pp. 117–126, 2008, Séminaire Maghrébin sur les Sciences et les Technologies de Séchage (SMSTS ‹08).
30. O. T. Adedoyin, A. Ojuawo, O. O. Adesiyun, F. Mark, and E. A. Anigilaje, "Poisoning due to yam flour consumption in five

families in Ilorin, Central Nigeria," West African Journal of Medicine, vol. 27, no. 1, pp. 41–43, 2008.

31. S. I. Adeleke, "Food poisoning due to yam flour consumption in Kano (Northwest) Nigeria," Online Journal of Health and Allied Sciences, vol. 8, no. 2, p. 10, 2009.
32. M. O. Oni and A. O. Omoniyi, "Studies on temperature influence on oviposition and development of immature stages of the yam beetle Dinoderus porcellus Lesne. Coleoptera: Bostrichidae on dried yam species," Journal of Agricultural Science, vol. 4, no. 2, pp. 213–218, 2012.
33. R. J. Bartelt and M. S. Hossain, "Chemical ecology of Carpophilus sap beetles (Coleoptera: Nitidulidae) and development of an environmentally friendly method of crop protection," Terrestrial Arthropod Reviews, vol. 3, no. 1, pp. 29–61, 2010.
34. J. M. Turaki, B. M. Sastawa, B. G. J. Kabir, and N. E. S. Lale, "Susceptibility of flours derived from various cereal grains to infestation by the rust-red flour beetle (Tribolium castaneum Herbst) (Coleoptera: Tenebrionidae) in different seasons," Journal of Plant Protection Research, vol. 47, no. 3, pp. 279–288, 2007.
35. N. E. S. Lale and B. A. Yusuf, "Insect pests infesting stored pearl millet Pennisetum glaucum (L.) R. Br. in Northeastern Nigeria and their damage potential," Cereal Research Communications, vol. 28, no. 1-2, pp. 181–186, 2000.
36. S. A. Babarinde, G. O. Babarinde, and O. Olasesan, "Physical and biophysical deterioration of stored plantain chips (Musa Sapientum L.) due to infestation of Tribolium castaneum herbst (Coleoptera: Tenebrionidae)," Journal of Plant Protection Research, vol. 50, no. 3, pp. 303–306, 2010.
37. L. Soldati, G. J. Kergoat, and F. L. Condamine, "Preliminary report on the Tenebrionidae (Insecta, Coleoptera) collected during the SANTO 2006 expedition to Vanuatu, with description of a new species of the genus Uloma Dejean, 1821," Zoosystema, vol. 34, no. 2, pp. 305–317, 2006.
38. P. Vernier, R. A. Dossou, and P. Letourmy, "La fabrication d›igname à partir de Dioscorea alata: influence de la variété et du type de chips sur le séchage, la conservation et les qualités organoleptiques,"African Journal of Root and Tuber Crops, vol. 3, pp. 62–67, 1999.

39. O. C. Eneh, "Enhancing Africa›s environmental management: integrated pest management for minimizing of agricultural pesticides pollution," Research Journal of Environmental Sciences, vol. 5, no. 6, pp. 521–529, 2011.
40. G. W. Norton, E. G. Rajotte, and V. Gapud, "Participatory research in integrated pest management: lessons from the IPM CRSP," Agriculture and Human Values, vol. 16, no. 4, pp. 431–439, 1999.
41. J. Brower and M. A. Mullen, "Effects of Xylocoris flavipes (Hemiptera: Anthocoridae) releases on moth populations in experimental peanut storages," Journal of Entomological Science, vol. 25, no. 2, pp. 268–276, 1990.
42. A. Russo, G. E. Cocuzza, and M. C. Vasta, "Life tables of Xylocoris flavipes (Hemiptera: Anthocoridae) feeding on Tribolium castaneum (Coleoptera: Tenebrionidae)," Journal of Stored Products Research, vol. 40, no. 1, pp. 103–112, 2004.
43. R. T. Arbogast and J. E. Throne, "Insect infestation of farm-stored maize in South Carolina: towards characterization of a habitat," Journal of Stored Products Research, vol. 33, no. 3, pp. 187–198, 1997.
44. J. Helbig, "Efficacy of Xylocoris flavipes (Reuter) (Het., Anthocoridae) to suppress Prostephanus truncatus (Horn) (Col., Bostrichidae) in traditional maize stores in Southern Togo," Journal of Applied Entomology, vol. 123, no. 8, pp. 503–508, 1999.
45. T. Imamura, M. Murata, and A. Miyanoshita, "Review biological aspects and predatory abilities of hemipterans attacking stored-product insects," Japan Agricultural Research Quarterly, vol. 42, no. 1, pp. 1–6, 2008.
46. K. T. Awadallah, A. I. Afifi, and I. I. El-Sebaey, "The biology of the reduviid, Allaeocranum biannulipes(Mon. et Sign.), a predator of stored product insect pests," Bulletin of the Entomological Society of Egypt, vol. 69, pp. 169–181, 1990.
47. D. P. Rees, "The effect of Teretriosoma nigrescens Lewis (Coleoptera: Histeridae) on three species of storage Bostrichidae infesting shelled maize," Journal of Stored Products Research, vol. 27, no. 1, pp. 83–86, 1991.

48. M. Camara, C. Borgemeister, R. H. Markham, and H. M. Poehling, "Electrophoretic analysis of the prey spectrum of Teretrius nigrescens (Lewis) (Col., Histeridae), a predator of Prostephanus truncatus (Horn) (Col., Bostrichidae), in Mexico, Honduras, and Benin," Journal of Applied Entomology, vol. 127, no. 6, pp. 360–368, 2003.
49. F. Dumaine, D. Dufour, C. Mestres, et al., "Effect of yam Dioscorea cayenenis-rotundata post-harvests treatments on yam chips quality," in Potential of Root Crops for Food and Industrial Resources. Twelfth Symposium of the International Society for Tropical Root Crops (ISTRC), September 10–16, 2000, Tsukuba, Japan, M. Nakatani and K. Komaki, Eds., vol. 12 of Symposium on Potential of Root Crops for Food and Industrial Resources, p. 394, Cultio Corporation, Tsukuba, Japon, 2002.
50. D. A. Akinpelu and E. M. Obuotor, "Antibacterial activity of Piliostigma thonningii stem bark,"Fitoterapia, vol. 71, no. 4, pp. 442–443, 2000.
51. S. A. M. Abdelgaleil, F. Hashinaga, and M. Nakatani, "Antifungal activity of limonoids from Khaya ivorensis," Pest Management Science, vol. 61, no. 2, pp. 186–190, 2005.
52. M. Nakatani, S. A. M. Abdelgaleil, S. M. I. Kassem et al., "Three new modified limonoids from Khaya senegalensis," Journal of Natural Products, vol. 65, no. 8, pp. 1219–1221, 2002.
53. D. Kubmarawa, M. E. Khan, A. M. Punah, and M. Hassan, "Phytochemical screening and antimicrobial efficacy of extracts from Khaya senegalensis against human pathogenic bacteria," African Journal of Biotechnology, vol. 7, no. 24, pp. 4563–4566, 2008.
54. M. Sale, N. De, J. H. Doughari, and M. S. Pukuma, "In vitro assessment of antibacterial activity of bark extracts of Khaya senegalensis," African Journal of Biotechnology, vol. 7, no. 19, pp. 3443–3446, 2008.V
55. S. A. Mitchell and M. H. Ahmad, "A review of medicinal plant research at the University of the West Indies, Jamaica, 1948–2001," West Indian Medical Journal, vol. 55, no. 4, pp. 243–269, 2006.
56. M. R. M. Ekué, B. Sinsin, O. Eyog-Matig, and R. Finkeldey, "Uses, traditional management, perception of variation and

preferences in ackee (Blighia sapida K.D. Koenig) fruit traits in Benin: implications for domestication and conservation," Journal of Ethnobiology and Ethnomedicine, vol. 6, article 12, 14 pages, 2010.

57. A. Khan, F. A. Gumbs, and A. Persad, "Pesticidal bioactivity of ackee (Blighia sapida Koenig) against three stored-product insect pests," Tropical Agriculture, vol. 79, no. 4, pp. 217–223, 2002.
58. A. Khan and F. A. Gumbs, "Repellent effect of ackee (Blighia sapida Koenig) component fruit parts against stored product insect pests," Tropical Agriculture, vol. 80, no. 1, pp. 19–27, 2003.
59. M. Tamiru, H. C. Becker, and B. L. Maass, "Diversity, distribution and management of yam landraces (Dioscorea spp.) in Southern Ethiopia," Genetic Resources and Crop Evolution, vol. 55, no. 1, pp. 115–131, 2008.
60. G. R. Kombo, A. Dansi, L. Y. Loko et al., "Diversity of cassava (Manihot esculenta Crantz) cultivars and its management in the department of Bouenza in the Republic of Congo," Genetic Resources and Crop Evolution, vol. 59, no. 8, pp. 1789–1803, 2012.
61. L. W. Kitch, O. Boukar, C. Endondo, and L. L. Murdock, "Farmer acceptability criteria in breeding cowpea," Experimental Agriculture, vol. 34, no. 4, pp. 475–486, 1998.
62. C. S. Gold, A. Kiggundu, A. M. K. Abera, and D. Karamura, "Selection criteria of Musa cultivars through a farmer participatory appraisal survey in Uganda," Experimental Agriculture, vol. 38, no. 1, pp. 29–38, 2002.
63. G. Abebe, T. Assefa, H. Harrun, T. Mesfine, and A. R. M. Al-Tawaha, "Participatory selection of drought tolerant maize varieties using mother and baby methodology: a case study in the semi-arid zones of the Central Rift Valley of Ethiopia," World Journal of Agricultural Sciences, vol. 1, no. 1, pp. 22–27, 2005.
64. G. Belay, H. Tefera, B. Tadesse, G. Metaferia, D. Jarra, and T. Tadesse, "Participatory variety selection in the Ethiopian cereal tef (Eragrostis tef)," Experimental Agriculture, vol. 42, no. 1, pp. 91–101, 2006.
65. A. Teshome, D. Patterson, Z. Asfew, J. K. Torrance, and J. T. Arnason, "Changes of Sorghum bicolorlandrace diversity and

farmers› selection criteria over space and time, Ethiopia," Genetic Resources and Crop Evolution, vol. 54, no. 6, pp. 1219–1233, 2007.

Chapter 4

Fumigation Toxicity of Essential Oil Monoterpenes to Callosobruchus Maculatus (Coleoptera: Chrysomelidae: Bruchinae)

Olufunmilayo E. Ajayi[1,2], Arthur G. Appel[1], and Henry Y. Fadamiro[1]

[1]Department of Entomology and Plant Pathology, Auburn University, 301 Funchess Hall, Auburn, AL 36849-5413, USA

[2]Department of Biology, Storage Technology Programme, Federal University of Technology, PMB 704, Akure 34001, Nigeria

ABSTRACT

The fumigant toxicity of eight essential oil components, 1-8-cineole, carvacrol, eugenol, (−)-menthone, (−)-linalool, S-(−)-limonene, (−)-β-pinene, and (+)-α-pinene, was tested against the cowpea weevil, Callosobruchus maculatus (Fabricius) (Coleoptera: Chrysomelidae), at 0.25–60 μL/L air doses. 1-8-Cineole, carvacrol, and eugenol caused complete adult mortality at 10 μL/L air 24 h after treatment. 1-8-Cineole and carvacrol were the most toxic with LD_{50} values of 0.24 and 0.6 μL/L air at 24 h, respectively. (−)-β-Pinene and (+)-α-pinene were the least toxic with LD_{50} values of 31 and 31.4 μL/L air at 24 h, respectively. Toxicity was negatively correlated with vapor pressure. 1-8-Cineole and carvacrol caused 100% oviposition deterrence at all doses tested. Eugenol and (−)-menthone completely inhibited adult emergence. S-(−)-Limonene, (−)-β-pinene, and (+)-α-pinene were not effective at preventing oviposition or adult emergence, suggesting that a lethal dose of the three oil components would be necessary to control C. maculatus infestations.

INTRODUCTION

Cowpea, Vigna unguiculata (L.) (Walp.), is an important food legume for millions of people throughout the semiarid regions of Africa, Asia, southern Europe, and North, Central, and South America [1]. The cowpea seed weevil, Callosobruchus maculatus (F.) (Coleoptera: Chrysomelidae: Bruchinae), is the major pest of stored cowpea seed in the tropics and subtropics due to the favorable climatic conditions [2, 3]. The insect infestscowpeas in the field and the subsequent population buildup in storage can cause complete weight loss of stored cowpeas within six months if no prophylactic measures are put in place [4]. Life history and development of this insect on cowpeas have been described in early studies [2, 5].

Stored grain pest control has relied mainly on synthetic chemicals most of which pose serious dangers to life and the environment. Many organochlorine and organophosphate chemicals are already banned, while the use of some is being restricted [6]. The use of methyl bromide, a major fumigant to control pests, is highly restricted [7]. Inhalation of phosphine gas, another grain fumigant, has been implicated in many human health problems [8]. Human health and environmental problems caused by synthetic pesticides, coupled with insects developing resistance to these chemicals, [9] have led to the need for alternative strategies that are ecofriendly and less prone to the development of resistance by pests [10]. Fumigation is still the most effective method of eliminating insect pests in stored products including cowpeas. The concern about the danger in the use of fumigants like methyl bromide and sulfuryl fluoride in pest control has shifted attention to evaluating essential oils as an effective and ecofriendly alternative [11].

An emerging pest control tactic for stored product pests is the use of essential oils. Essential oils are composed of complex mixtures of monoterpenes, biogenetically related phenols, and sesquiterpenes obtained from plants through steam distillation [12].

Many studies have demonstrated contact and fumigant toxicity of plant essential oils and their components to several species of stored product insects at different life stages [13, 14]. Suthisut et al. [15] reported on the toxicity of essential oils of Alpinia conchigera (Griff), Zingiber zerumbet (L.) Roscoe ex Sm, Curcuma zedoaria (Christm.) Roscoe, and their components to Sitophilus zeamais (L.) and Tribolium castaneum (Herbst.). Various essential oils havealso been tested against C. maculatus [16–18]. For example, Hura crepitans (L.) seed oil was reported as toxic to adult and immature stages of C. maculatus [16]. Similarly, oil of Ziziphora clinopodioides (Boiss) caused significant adult and egg mortality [17]. The majority of the above studies on the toxicity of essential oils againstC. maculatus tested crude essential oils rather than individual synthetic compounds, making it difficult to attribute toxicity or oviposition deterrence to particular components. Furthermore, most of the studies did not determine the fate of eggs laid during fumigation (i.e., larval development and adult emergence from eggs laid during fumigation with the essential oils).

The objective of this study was to test the fumigant toxicity of some selected essential oil components at low concentrations against C. maculatus. The essential oil components tested in this study were chosen because they are major constituents of essential oils that have been found to have high pesticidal properties against store product pests, and the pure forms have also been reported toxic to some stored grain pests [12, 19–21]. Many essential oil components are known neurotoxins [20, 22–24]. We hypothesized that fumigation with low concentrations of the essential oil components will affect the nervous system of C. maculatus and thus prevent egg laying on stored seeds. A second hypothesis was that the efficacy of the essential oil components as fumigants is related to their physical properties. First, the fumigant toxicity of the oil components to adult C. maculatuswas determined in fumigation chambers. The ability of females to lay viable eggs during fumigation was then assessed using the number of eggs laid. The fate of the eggs laid during fumigation was also determined by recording adult emergence. Finally, toxicity of the essential oil components was correlated with some physical properties of the oil components to identify the factors that best predict toxicity.

MATERIALS AND METHODS

Eight essential oil components, 1-8-cineole, carvacrol, eugenol, (−)-menthone, (−)-linalool, S-(−)-limonene, (−)-β-pinene, and (+)-α-pinene, obtained from Sigma-Aldrich (St. Louis, MO) were tested in the study. The purity of the oil components ranged between 95 and 99%.

Cowpea Seed and Insect Culture

Cowpea (cultivar California black-eye) seeds were purchased from a local grocery store in Auburn, AL, USA. The seeds were kept in polythene bags, sealed and kept in a freezer maintained at −20°C for 72 hours until they were used for insect cultures or the bioassays [25].

A fresh culture of C. maculatus was started from a colony that had been maintained in the laboratory for over 50 generations by placing 25 pairs of 2-day-old male and female beetles in 1-L wide-mouthed glass Mason jars containing 100 g of cowpea seeds. The jars were placed in a rearing chamber maintained at 28 ± 2°C, 70 ± 5% r.h, and 12 : 12 h photoperiod [16, 26]. The beetles were sexed using the keys described by Rees [27]. Female beetles were allowed to lay eggs on the seeds for 24 hours after which they were removed. The seeds containing eggs were kept in a rearing chamber until adult emergence.

Fumigation with Essential Oil Components

The fumigant toxicity of eight essential oil components at 0.25, 5, 10, 20, 40, and 60 μL/L was investigated by exposing five pairs of 1-day-old adult male and female C. maculatus to vapors from the oil components in 1 L wide-mouthed glass Mason jars containing 20 g of cowpea seeds. Jars without essential oil components served as the controls. Each treatment was replicated four times. The fumigation lasted for 24 hours at abovementioned environmental conditions. Data on adult mortality was recorded as a percentage of the total number of the beetles fumigated. Data were arcsine transformed and analyzed using one-way analysis of variance (ANOVA). Tukey's HSD test was used to separate the means. Regression analysis was used to estimate LD_{50} in the adult mortality tests [28]. Stepwise multiple linear regression was used to reveal the physical property of the oil components that predicts LD_{50}.

To determine the effects of fumigant toxicity of the essential oil components on oviposition deterrence, the number of eggs on all cowpea seeds in each replicate was recorded after fumigation test. The egg-bearing seeds were returned back to the jars and placed in the rearing chamber undisturbed for four weeks after which they were examined for adult emergence. Adult emergence was monitored daily for 10 days starting from 30 days after treatment. Data on adult emergence were used to determine the fate of the eggs laid during fumigation. Logarithmic transformation was performed on oviposition deterrence and adult emergence data. ANOVA and Tukey's HSD test were used to compare the number of eggs laid and number of adults that emerged. All statistics were done using SPSS at $\alpha = 0.05$.

RESULTS

Fumigant Toxicity to Adults

No mortality was observed in untreated controls during this study. The effect of the tested eight essential oil components on the survival of adult C. maculatus was concentration dependent. Three essential oil components, 1-8-cineole, carvacrol, and eugenol, caused ≥90% mortality to adult beetles at doses of as low as 5 μL/L of air within 24 hours. (−)-Menthone, (−)-linalool, S-(−)-limonene, and the two structural isomers of pinene also caused 100% mortality to adult beetles, but at significantly greater doses (20 μL/L and above) (Figure 1).

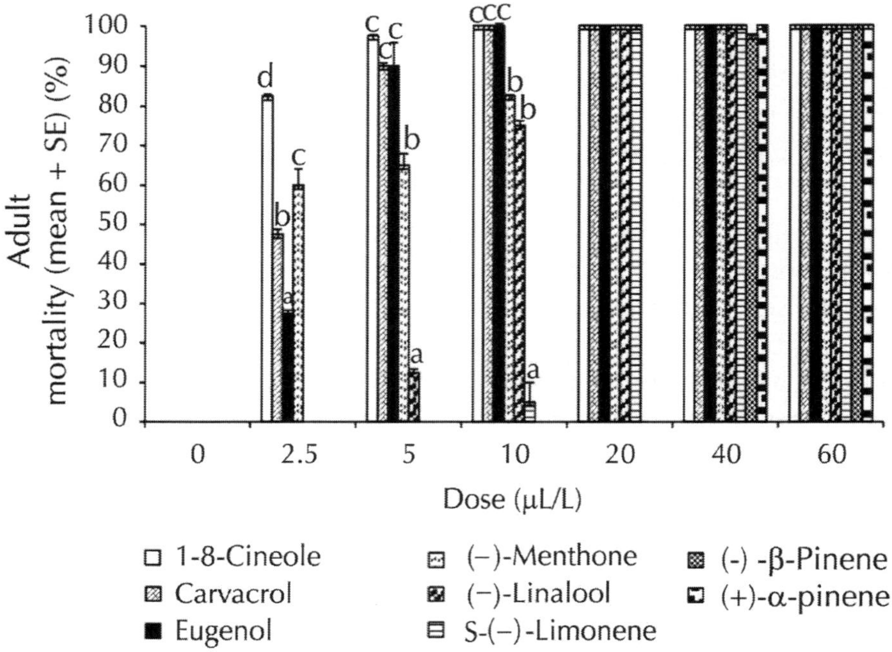

Figure 1: Mortality of adult C. maculatus exposed to eight essential oil monoterpenes.

The LD_{50} values obtained in this study (Figure 2) were step-wisely regressed with the density, boiling point, and vapor pressure of the oil components. The analysis revealed vapor pressure as the physical property of the essential oil components that best predicts the LD_{50} ($F_{1,30} = 72.57, P=0.001$) and accounted for over 70% of the variances. In general, toxicity was negatively correlated with vapor pressure (Figure 3). Based on the regression analysis, the predicted LD_{50} of essential oils can be calculated as follows:

$$\text{Predicted } LC_{50} = 13.044 \times \text{vapor pressure} + 3.081. \quad (1)$$

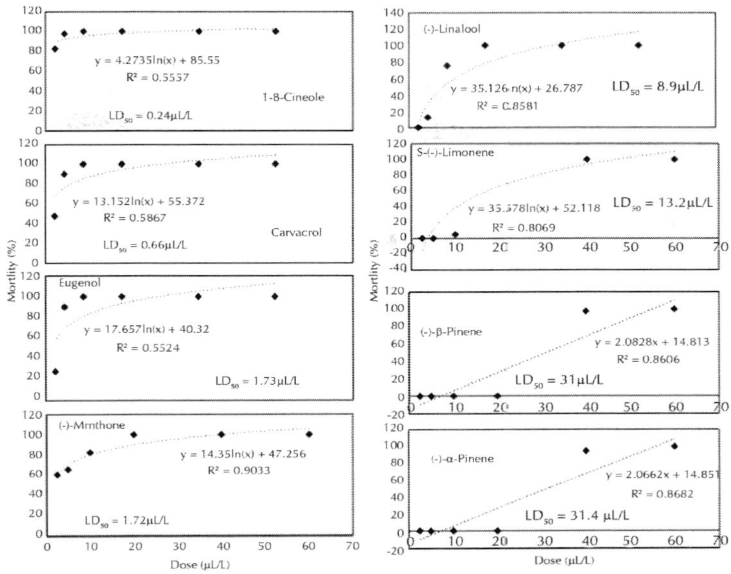

Figure 2: Dose-response curves of eight monoterpenes tested against C. maculatus.

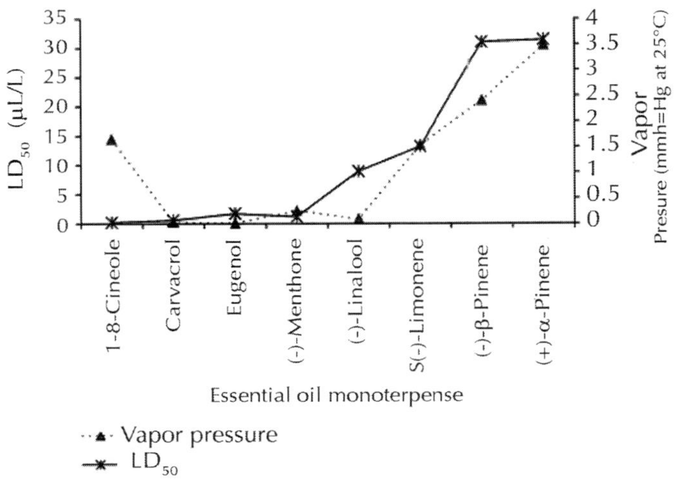

Figure 3: Relationship between LD_{50} and vapor pressure of eight essential oil monoterpenes tested against C. maculatus.

Oviposition Deterrence and Adult Emergence

1-8-Cineole and carvacrol completely deterred oviposition at all doses tested (Figure 4). Numbers of eggs laid decreased with increasing dose for eugenol, (−)-menthone (−)-linalool, S-(−)-limonene, and the two structural isomers of pinene. For example, eugenol-treated seeds had an average of 18.25 and 6.25 eggs at 2.5 and 5.0 μL/L, respectively. The two isomers of pinene had significantly higher numbers of eggs compared to (−)-linalool and S-(−)-limonene. All eight essential oil components caused 100% oviposition deterrence at the highest dose tested (60 μL/L) (Figure 4).

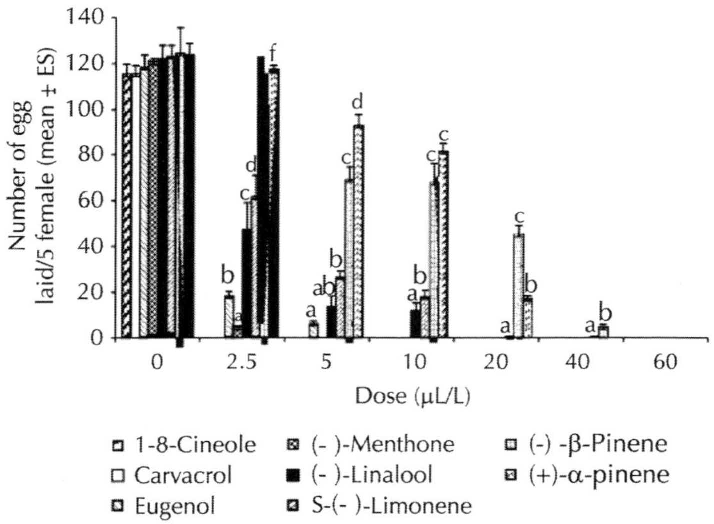

Figure 4: Number of eggs laid by C. maculatus during fumigation with essential oil monoterpenes. ANOVA and Tukey's HSD compared monoterpenes at each dose tested.

Among the oil treatments which did not completely prevent oviposition, eugenol and (−)-menthone completely inhibited adult emergence at all doses tested (Figure 5). Adult emergence decreased with increasing dose of (−)-linalool, S-(−)-limonene, and the two pinene isomers (Figure 5). Adult emergence in seeds treated with (+)-α-pinene and (−)-β-pinene was significantly (P=0.001) higher than in linalool and S-(−)-limonene, at 0.25–20 μL/L air (Figure 5).

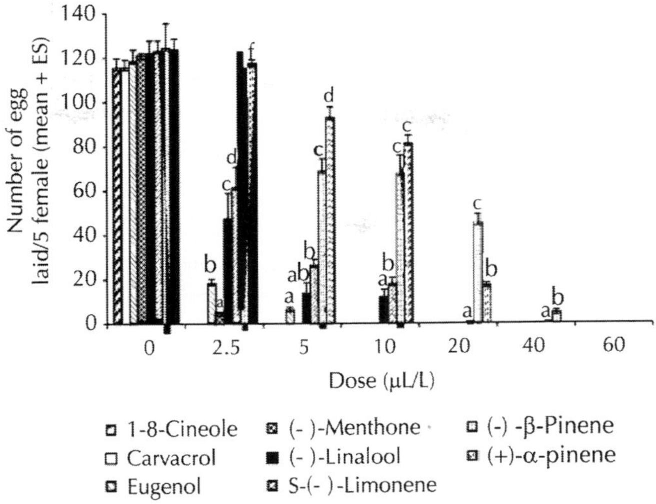

Figure 5: Number of adult C. maculatus that emerged from eggs laid during fumigation with essential oil monoterpenes. ANOVA and Tukey's HSD test compared monoterpenes at each dose tested.

DISCUSSION

Fumigant Toxicity to Adults

The results of this study demonstrated fumigant toxicity of some essential oil components to C. maculatus. Three compounds, 1-8-cineole, carvacrol, and eugenol, caused almost 100% adult mortality at doses as low as 5 μL/L air. Similar levels of adult beetle mortality were achieved with the remaining five components, (−)-linalool,S-(−)-limonene, (−)-menthone, and the two structural isomers of pinene, but only at relatively greater concentrations (≥10 μL/L air).

Previous studies have reported on the toxicity of essential oils to C. maculatus and other stored product insect pests [21, 29–31]. The toxicity of essential oils tested was attributed to the components of the oils [20, 32]. Fahimeh et al. [29] reported toxicity of essential oil of Zataria multiflora (Boiss) (Lamiaceae) to C. maculatus at doses as low as 9 μL/L air 24 hours after treatment. It is worth mentioning that Zataria multiflora is a major source of carvacrol, which was also toxic to adult C. maculatus causing 100% mortality at 10 μL/L air in the present study. However, our LD_{50} for carvacrol (0.6 μL/L air) (Figure 2) was about one-fourteenth of the LD_{50}value (8.81 μL/L air) reported by Fahimeh et al. [29]. Sharifian et al. [33] reported that oil extracts of Mugwort (Artemisia vulgaris L.), a major source of (+)-α-pinene, were effective against C. maculatus. The estimated LD_{50} of 53 μL/L air at 24 hours was greater than the LD_{50} of 31.4 μL/L air at 24 hours obtained in our study using the neat form of (+)-α-pinene. Arezo et al. [34] estimated LD_{50} of 24 μL/L air for (−)-linalool at 24 hours which was almost three times greater than the LD_{50} value of 8.9 μL/L air obtained for (−)-linalool in this present study. Similarly, the 13.2 μL/L air we obtained for S-(−)-limonene was eight times less than the LD_{50} value of 106 μL/L air obtained by Ramin and Yunes [31] with essential oils from the peels of Citrus sinensis (L.). Synthetic (pure) monoterpenes were tested in the present study and that may explain the much lower LD_{50} values compared to that reported in the earlier studies with whole essential oils. Furthermore, it is plausible that other components of essential oils may have interacted to reduce efficacy of the monoterpenes in the whole oils tested.

Earlier studies have also revealed toxicities of the monoterpenes we tested in this study to other stored grain pests [35–38]. Carvacrol and 1‥8-cineole were found toxic to Tenebrio molitor (F.) at LD_{50} of 5.52 and 5.71 μL, respectively [36]. Limonene had varied level of toxicity to Rhyzopertha dominica (F.), Sitophilus oryzae (L.), andTribolium castaneum (Herbst.) [35]. Ogendo et al. [37] reported that mortalities of S. oryzae, R. dominica, T. castaneum, Oryzaephilus surinamensis (L.), and Callosobruchus chinensis (F.) with α- and β-pinenes ranged between 7.3–100% and 32.5–100%, respectively. (−)-Linalool and (−)-menthone were also found to cause 100% mortality of S. zeamais (L.) at 30 μL/μg [38]. Source and purity may affect bioactivity of essential oils and the monoterpenes, and also age, size, sex, insect order, and species influence insects' susceptibility to chemicals [34]. Any of the above-mentioned factors or a combination of two or more of these factors may have accounted for the differences between the results obtained in the present study and the previous related studies.

Oviposition Deterrence and Adult Emergence

Two of the eight treatments, 1-8-cineole and carvacrol, completely deterred oviposition. (−)-Menthone and eugenol treated seeds had significantly fewer eggs while seeds treated with (+)-α-pinene and (−)-β-pinene had the highest numbers of eggs. Among the six treatments which did not cause complete prevention of oviposition, eugenol and (−)-menthone completely inhibited adult emergence. The two pinene isomers had greater numbers of adult emergence. Previous related studies have tested the toxicity of crude essential oils to eggs of C. maculatuslaid during fumigation [39, 40], but we are not aware of any studies that specifically tested synthetic (pure) essential oil components. Eman and Abass [40], in their study of efficacy of five essential oils against C. maculatus, reported significant reduction in oviposition with 1% acetone solution of Mentha rotundifolia (L.) andMentha pulegium (L.). M. rotundifolia oil contains a low percentage (8.5%) of 1-8-cineole, while M. pulegium oil has 0.05, 0.2, and 0.4% of (+)-α-pinene, (−)-β-pinene, and S-(−)-limonene, respectively. Interestingly, S-(−)-limonene and the two isomers of pinene were not effective in the present study as oviposition deterrent or in preventing adult emergence even at high doses. Oil of Elettaria cardamomum (L.), a major source of 1-8-cineole (55.7%), also reduced oviposition by C. maculatus [40]. Other components of the oil may have interacted to lower the efficacy of 1-8-cineole content. Essential oils and their components possess low or no toxicity against vertebrates [20, 41, 42]. Acute toxicity (LD_{50}) of 1-8-cineole (2,480 mg/kg BW), eugenol (2,680 mg/kg BW), and linalool (>1000 mg/kg BW) in rats is far less than that of some approved plant-based pesticides like pyrethrins (300–350 mg/kg BW) and pyrethrum (1500 mg/kg BW) [20, 32, 43]. Besides, they are easily biodegradable; hence, the use of these oil components to protect cowpea seeds may not constitute problems to man and environment.

In summary, our results showed that 1-8-cineole, (−)-menthone, carvacrol, and eugenol are highly toxic to adult and egg stages of C maculatus, even at low concentrations. Vapor pressure was the best predictor of toxicity of the essential oil components. While 1-8-cineole and carvacrol completely prevented oviposition, some eggs were laid on seeds treated with eugenol and (−)-menthone but the eggs did not develop to adults. These suggest that 1-8-cineole, carvacrol, eugenol, or (−)-menthone can be used to prevent population build-up of C. maculatus in stored cowpea. The results also showed that fumigation with a single lethal dose or repeated treatment with (−)-linalool, S-(−)-limonene, (+)-α-pinene, and (−)-β-pinene may be necessary for effective control of C. maculatus in stored cowpea. Essential oil components may also be combined with modified atmosphere or other tactics such as Purdue Improved Bag to achieve effective protection of stored cowpea.

ACKNOWLEDGMENTS

The authors would like to acknowledge The Schlumberger Foundation Inc., Houston, Texas, for supporting this research through a fellowship to OEA. The authors thank Marla J. Eva for laboratory assistance.

REFERENCES

1. B. B. Singh, H. A. Ajeigbe, S. A. Tarawali, S. Fernandez-Rivera, and M. Abubakar, "Improving the production and utilization of cowpea as food and fodder," Field Crops Research, vol. 84, no. 1-2, pp. 169–177, 2003.
2. S. R. Singh, L. E. N. Jackai, J. H. R. Dos Santo, and C. B. Adalla, "Insect pest of Cowpea," in Insect Pest of Food Legumes, S. R. Singh, Ed., pp. 43–80, John Wiley & Sons, Chichester, UK, 1990.
3. N. Z. Dimetry, S. El-Gengaihi, and A. M. E. Abd El-Salam, "Protection of stored cowpea fromCallosobruchus maculatus (F.) Attack by some plant extracts formulations in different storage sacks,"Herba Polonica, vol. 53, pp. 71–84, 2007.
4. A. Sanon, L. C. B. Dabiré, A. P. Ouédraogo, and J. Huignard, "Field occurrence of bruchid pests of cowpea and associated parasitoids in a sub humid zone of Burkina Faso: importance on theinfestation of two cowpea varieties at harvest," Journal of Plant Pathology, vol. 4, no. 1, pp. 14–20, 2005.
5. P. A. Edde and C. I. Amatobi, "Seed coat has no value in protecting cowpea seed against attack byCallosobruchus maculatus (F.)," Journal of Stored Products Research, vol. 39, no. 1, pp. 1–10, 2003

6. G. Forget, "Balancing the need for pesticides with the risk to human health," in Impact of Pesticide Use on Health in Developing Countries, G. Forget, T. Goodman, and A. de Villiers, Eds., vol. 2, International Development Research Centre, Ottawa, Canada, 1993.
7. R. Meadows, "News overview: researchers develop alternatives to methyl bromide fumigation," California Agriculture, vol. 67, no. 3, pp. 125–127, 2013.
8. US Environmental Protection Agency (US EPA), Phosphine. Integrated Risk Information Service (IRIS), US EPA, Washington, DC, USA, 2003.
9. M. K. Mahmud, M. M. H. Khan, M. Husain, M. I. Alam, and M. S. I. Afrad, "Toxic effects of different plant oils on pulse beetle Callosobruchus chinensis Linn. (Coleoptera: Bruchidae)," Journal of Asiatic Society of Bangladesh: Science, vol. 28, no. 1, pp. 11–18, 2002.
10. F. E. Dayan, C. L. Cantrell, and S. O. Duke, "Natural products in crop protection," Bioorganic and Medicinal Chemistry, vol. 17, no. 12, pp. 4022–4034, 2009
11. M.-J. Sim, D.-R. Choi, and Y.-J. Ahn, "Vapor phase toxicity of plant essential oils to Cadra cautella(Lepidoptera: Pyralidae)," Journal of Economic Entomology, vol. 99, no. 2, pp. 593–598, 2006.
12. M. Isman, C. Machial, S. Miresmailli, and L. Bainard, "Essential oil-based pesticides: new insights from old chemistry," in Pesticides Chemistry, H. Ohkawa, H. Miyagawa, and P. Lee, Eds., pp. 201–209, Wiley-VCH, Weinheim, Germany, 2007.
13. D. P. Papachristos and D. C. Stamopoulos, "Selection of Acanthoscelides obtectus (Say) for resistance to lavender essential oil vapour," Journal of Stored Products Research, vol. 39, no. 4, pp. 433–441, 2003.
14. U. R. Pathipati, "Fumigant and contact toxic potential of essential oils from plant extracts against stored product pests," Journal of Biopesticides, vol. 5, no. 2, pp. 120–128, 2012.
15. D. Suthisut, P. G. Fields, and A. Chandrapatya, "Fumigant toxicity of essential oils from three Thai plants (Zingiberaceae) and their major compounds against Sitophilus zeamais, Tribolium castaneum and two parasitoids," Journal of Stored Products Research, vol. 47, no. 3, pp. 222–230, 2011.

16. C. O. Adedire and O. E. Ajayi, "Potential of sandbox, Hura crepitans L. seed oil for protection of cowpea seeds from Callosobruchus maculatus (F.) (Coleoptera: Bruchidae) infestation," Journal of Plant Diseases and Protection, vol. 110, no. 6, pp. 602–610, 2003.
17. F. A. Lolestani and N. Shayesteh, "Fumigant toxicity of Ziziphora clinopodioides (Boiss.) (Lamiaceae) against adults and eggs of Callosobruchus maculatus (Fab.) (Coleoptera: Bruchidae)," Journal of Biological Sciences, vol. 9, no. 1, pp. 92–95, 2009.
18. M. E. A. Ahmed, "Fumigant toxicity of seven essential oils against the cowpea weevil, Callosobruchus maculatus (F.) and the rice weevil, Sitophilus oryzae (L.)," Egyptian Academic Journal of Biological Sciences, vol. 2, pp. 1–6, 2010.
19. S. Lee, R. Tsao, C. Peterson, and J. R. Coats, "Insecticidal activity of monoterpenoids to western corn rootworm (Coleoptera: Chrysomelidae), twospotted spider mite (Acari: Tetranychidae), and house fly (Diptera: Muscidae)," Journal of Economic Entomology, vol. 90, no. 4, pp. 883–892, 1997.
20. O. Koul, S. Walia, and G. S. Dhaliwal, "Essential oils as green pesticides: potential and constraints,"Pesticide International, vol. 4, no. 1, pp. 63–84, 2008.
21. A. K. Phillips and A. G. Appel, "Fumigant toxicity of essential oils to the German cockroach (Dictyoptera: Blattellidae)," Journal of Economic Entomology, vol. 103, no. 3, pp. 781–790, 2010.
22. M. B. Isman, "Plant essential oils for pest and disease management," Crop Protection, vol. 19, no. 8–10, pp. 603–608, 2000.
23. E. Enan, "Insecticidal activity of essential oils: octopaminergic sites of action," Comparative Biochemistry and Physiology C: Toxicology and Pharmacology, vol. 130, no. 3, pp. 325–337, 2001.
24. C. Regnault-Roger, C. Vincent, and J. T. Arnason, "Essential oils in insect control: low-risk products in a high-stakes world," Annual Review of Entomology, vol. 57, pp. 405–424, 2012.
25. J. A. Johnson and K. A. Valero, "Use of commercial freezers to control cowpea weevil, Callosobruchus maculatus (Coleoptera: Bruchidae), in organic garbanzo beans," Journal of Economic Entomology, vol. 96, no. 6, pp. 1952–1957, 2003.

26. E. O. Onagbola and H. Y. Fadamiro, "Electroantennogram and behavioral responses of Pteromalus cerealellae to odor stimuli associated with its host, Callosobruchus maculatus," Journal of Stored Products Research, vol. 47, no. 2, pp. 123–129, 2011.
27. D. Rees, Insects of Stored Products, CIRO Publishing, Collingwood, Australia, 2004.
28. SPSS, Statistical Package for Social Sciences, 16.0 for windows. SPSS Inc, Chicago, Ill, USA, 2008.
29. R. Fahimeh, M. Saeid, S. Mahmood, and A. Habib, "Chemical composition and insecticidal activity of essential oil of Zataria multiflora Boiss. (Lamiaceae) against Callosobruchus maculatus (F.) (Coleoptera : Bruchidae)," IOBC-WPRS Bulletins, vol. 69, pp. 281–288, 2011.
30. V. Rozman, I. Kalinovic, and Z. Korunic, "Toxicity of naturally occurring compounds of Lamiaceae and Lauraceae to three stored-product insects," Journal of Stored Products Research, vol. 43, no. 4, pp. 349–355, 2007.
31. T. Ramin and K. Yunes, "Evaluation of fumigant toxicity of orange peel Citrus sinensis (L.) essential oil against three stored product insects in laboratory condition," Munis Entomology and Zoology, vol. 7, pp. 352–358, 2012.
32. D. R. Batish, H. P. Singh, R. K. Kohli, and S. Kaur, "Eucalyptus essential oil as a natural pesticide," Forest Ecology and Management, vol. 256, no. 12, pp. 2166–2174, 2008.
33. I. Sharifian, S. M. Hashemi, and A. Darvishzadeh, "Fumigant toxicity of essential oil of Mugwort (Artemisia vulgaris L.) against three major stored product beetles," Archives of Phytopathology and Plant Protection, vol. 46, no. 4, pp. 445–450, 2013
34. A. Davoudi, N. Shayesteh, D. Shirdel, and A. Hosseinzadeh, "Effect of diethyl maleate on toxicity of linalool against two stored product insects in laboratory condition," African Journal of Biotechnology, vol. 10, no. 48, pp. 9918–9921, 2011.
35. A. K. Tripathi, V. Prajapati, S. P. S. Khanuja, and S. Kumar, "Effect of d-limonene on three stored-product beetles," Journal of Economic Entomology, vol. 96, no. 3, pp. 990–995, 2003.
36. R. K. Lima, M. D. G. Cardoso, J. C. Moraes, S. M. Carvalho, V. G. Rodrigues, and L. G. L. Guimarães, "Chemical composition

and fumigant effect of essential oil of Lippia sidoides cham. and monoterpenes against Tenebrio molitor (L.) (Coleoptera: Tenebrionidae)," Ciência e Agrotecnologia, vol. 35, no. 4, pp. 664–671, 2011.

37. J. O. Ogendo, A. L. Deng, M. Kostyukovsky, et al., "Fumigant toxicity of five essential oil constituents against major stored-product insect pests of food grains. Research and application summary," inProceedings of the 2nd RUFORUM Biennial Meeting, Entebbe, Uganda, September 2010.
38. E. Yildirim, B. Emsen, and S. Kordali, "Insecticidal effects of monoterpenes on Sitophilus zeamaisMotschulsky (Coleoptera: Curculionidae)," Journal of Applied Botany and Food Quality, vol. 86, pp. 198–204, 2013
39. E. A. Eman and M. H. Abass, "Chemical composition and efficiency of five essential oils against the pulse beetle Callosobruchus maculatus (F.) on Vigna radiata seeds," The American-Eurasian Journal of Agriculture and Environmental Science, vol. 8, pp. 411–419, 2010.
40. H. Abbasipour, M. Mahmoudvand, F. Rastegar, and M. H. Hosseinpour, "Fumigant toxicity and oviposition deterrency of the essential oil from cardamom, Elettaria cardamomum, against three stored—product insects," Journal of Insect Science, vol. 11, article 165, 2011.
41. E. Enan, M. Beigler, and A. Kende, "Insecticidal action of terpenes and phenols to cockroaches: effect on octopamine receptors," in Proceedings of the International Symposium on Plant Protection, Gent, Belgium, 1998.
42. F. Bakkali, S. Averbeck, D. Averbeck, and M. Idaomar, "Biological effects of essential oils—a review," Food and Chemical Toxicology, vol. 46, no. 2, pp. 446–475, 2008.
43. J. E. Casida and G. B. Quistad, Pyrethrum Flowers: Production, Chemistry, Toxicology and Uses, Oxford University Press, Oxford, UK, 1995.

Chapter 5

Pseudo-Backcrossing Design for Rapidly Pyramiding Multiple Traits into a Preferential Rice Variety

Siriphat Ruengphayak[1,2], Ekawat Chaichumpoo[1], Supaporn Phromphan[1], Wintai Kamolsukyunyong[3], Wissarut Sukhaket[1], Ekapol Phuvanartnarubal[1], Siripar Korinsak[3], Siriporn Korinsak[3], and Apichart Vanavichit[1,3,4]

[1]Rice Science Center, Kasetsart University, Kamphaeng Saen 73140, Nakhon Pathom, Thailand

[2]Interdisciplinary Graduate Program in Genetic Engineering, Kasetsart University, Chatuchak 10900, Bangkok, Thailand

[3]Rice Gene Discovery, National Center for Genetic Engineering and Biotechnology (BIOTEC) National Science and Technology Development Agency (NSTDA), Kasetsart University, Kamphaengsaen 73140, Nakhon Pathom, Thailand

[4]Agronomy Department, Faculty of Agriculture at Kamphaeng Saen, Kasetsart University, Kamphaeng Saen 73140, Nakhon Pathom, Thailand

ABSTRACT

Background

Pyramiding multiple genes into a desirable genetic background can take years to accomplish. In this paper, a pseudo-backcrossing scheme was designed to shorten the backcrossing cycle needed. PinK3, an aromatic and potentially high-yielding rice variety—although one that is intolerant to flash flooding (Sub) and susceptible to bacterial leaf blight (BB), leaf-neck blast (BL) and the brown planthopper (BPH)—was used as a genetic basis for significant improvements through gene pyramiding.

Results

Four resistance donors with five target genes (Sub1A-C, xa5, Xa21, TPS and SSIIa) and three QTLs (qBph3, qBL1 and qBL11) were backcrossed individually using markers into the pseudo-recurrent parent 'PinK3' via one cycle of backcrossing followed by two cycles of pseudo-backcrossing and three selfings with rigorous foreground marker-assisted selection. In total, 29 pseudo-backcross inbred lines (BILs) were developed. Genome composition was surveyed using 61 simple sequence repeats (SSRs), 35 of which were located on six carrier chromosomes, with the remainder located on six non-carrier chromosomes. The recurrent genome content (%RGC) and donor genome content (%DGC), which were based on the physical positions of BC_1F_2, ranged from 69.99 to 88.98% and 11.02 to 30.01%, respectively. For the pseudo-BC_3F_3BILs, the %RGC and %DGC ranged from 74.50 to 81.30% and 18.70 to 25.50%, respectively. These results indicated that without direct background selection, no further increases in %RGC were obtained during pseudo-backcrossing, whereas rigorous foreground marker-assisted selection tended to reduce linkage drag during pseudo-backcrossing. The evaluation of new traits in selected pseudo-BC_3F_3BILs indicated significant improvements in resistance to BB, BL, BPH and Sub compared with PinK3, as well as significant improvements in grain yield (21-68%) over the donors, although yield was 7-26% lower than in 'PinK3'. All pyramided lines were aromatic and exhibited improved starch profiles, rendering them suitable for industrial food applications.

Conclusions

Results show that our new pyramiding platform, which is based on marker-assisted pseudo-backcrossing, can fix five target genes and three QTLs into a high-yielding pseudo-recurrent background within seven breeding cycles in four years. This multiple pseudo-backcrossing platform decreases the time required to generate new rice varieties exhibiting complex, durable resistance to biotic and abiotic stresses in backgrounds with desirable qualities.

BACKGROUND

Rice (Oryza sativa L.) is an important staple food crop and a major part of the diet of more than half of the world's population. Approximately 90% of rice is grown in Asia. Worldwide, approximately 79 million ha of irrigated lowland rice provides 75% of the world's rice production (Maclean et al. [2002]; Bouman et al. [2007]). Therefore, irrigated rice remains the most important production system for maintaining food security, particularly in Asian countries. Rice production in irrigated areas of Thailand has been frequently and strongly affected by abiotic stresses resulting from unfavorable climatic changes, such as flooding and drought, as well as by biotic stresses caused by bacterial leaf blight (BB), leaf/neck blast (BL) and brown planthopper (BPH). Therefore, new successful breeding lines must possess multiple types of resistance to both biotic and abiotic stresses, as well as demonstrating specific grain qualities and high yield.

Three popular breeding methods used for gene pyramiding are pedigree, backcrossing and recurrent selection. In cross-pollinating crops, gene pyramiding is accomplished through recurrent selection. Successful quantitative trait locus (QTL) introgression depends on the optimized expression of newly introgressed QTLs in the recipient genome background, with the aim of maximizing productivity. A general framework for addressing these considerations through the pyramiding of multiple QTLs into a single favorable genetic background has been proposed, although such techniques are time-consuming (Servin et al. [2004]).

To pyramid several new QTLs, stepwise crossing schemes can be designed, although such schemes can require many generations of breeding. If the ultimate goal is to improve a specific desired variety, parallel backcrossing of single donors can most effectively be carried out using both foreground and background selection. Recently, successful gene/QTL pyramiding programs were reported in Thai Jasmine (Win et al. [2012]; Luo and Yin, [2013]), Basmati (Singh et al.[2012]; Singh et al. [2013]), Koshihikari (Ashikari and Matsuoka, [2006]; Tomita, [2009]), Zhenshan 97 (Wang et al. [2012]) and 93–11 (Zong et al. [2012]) rice varieties. However, when the target traits are quantitatively controlled, combining several QTLs can take years to accomplish. For multiple QTL pyramiding, three phases are necessary: the creation of near-isogenic lines (NILs), genotype assembly, and the extraction of pure lines. Specific backcrossed recombinant inbred lines (RILs) carrying four main-effect QTLs and four epistatic-effect QTLs were pyramided into the elite cultivar Zhenshan97 (Wang et al. [2012]). To combine greater numbers of QTLs, marker-assisted phenotypic selection (MAS) has been developed, which is a novel approach for QTL pyramiding of up to 24 QTLs from a single crossing (Zong et al. [2012]). QTL pyramiding via NILs was successfully used to improve disease and lodging resistance, as well as to increase the harvest index (Luo and Yin, [2013]). However, all of these approaches require many years to complete.

To shorten the backcross breeding cycle, we propose a modified form of pseudo-backcrossing. The original design of pseudo-backcrossing originated from tree breeding methods, where F_1 plants resulting from a single cross are backcrossed to alternate recurrent parents to avoid inbreeding depression (Bouquet [1986]). Pseudo-backcrossing is commonly used in perennial plants, including grape (Molnár et al. [2007]), eucalyptus (Kullan et al. [2012]), poplar (Novaes et al. [2009]) and oil palm (Montoya et al. [2014]). For multiple gene/QTL pyramiding, several donors are used to create newly improved genotypes; however, the need to maintain the preferred genetic background is equally important. Therefore, the introduction of pseudo-backcrossing could benefit multiplex gene pyramiding. Most rice varieties are not tolerant to flash flooding. Submergence tolerance is determined by the major Sub1QTL on chromosome 9 with relatively high heritability. The new

Thai jasmine rice variety 'Khao Dawk Mali 105 (KDML105)' was developed for submergence (Sub) tolerance using marker-assisted backcross breeding (MAB) and has been released to farmers (Siangliw et al.[2003]). Growing rice plants also suffer from epidemic diseases such as rice blast (BL) and bacterial leaf blight (BB). Rice BL, which is caused by the fungal pathogen Pyricularia oryzae(teleomorph: Magnaporthe oryzae), is a major rice disease in irrigated rice-growing areas worldwide (Ou [1985]). QTLs for broad-spectrum resistance to rice BL have been reported on chromosomes 1 and 11 from JHN and on chromosomes 2 and 12 from IR64 (Sirithunya et al.[2002]; Noenplab et al. [2006]; Sreewongchai et al. [2010]). Major effective QTLs for Thai blast isolates were located on chromosomes 1 and 11, which are flanked by RM212 and RM319 on qBL1 and by RM224-RM144 on qBL11 (Noenplab et al. [2006]; Wongsaprom et al. [2010]). The presence of qBL1 and qBL11 have strong effects on blast resistance in Thailand. BB, which is caused by the bacterium Xanthomonas oryzae pv oryzae can be effectively controlled using resistant varieties. Several resistance genes, including Xa4, xa5 and Xa21, are effective sources of resistance for marker-assisted gene pyramiding in rice (Korinsak et al. [2009b]; Suh et al. [2013]).

Among insect rice pests, the brown planthopper (BPH, Nilaparvata lugens Sta° l), is considered one of the most serious pests of irrigated rice. The damage caused by BPH feeding has a major effect on crop growth and yield (Watanabe and Kitagawa, [2000]; Yuan et al. [2005]). BPH not only feeds on the rice plant directly but also transmits viruses that cause severe diseases (Heinrichs [1979]). The use of BPH resistance genes has been recognized as the most economic, effective and environmentally friendly solution to this problem. The stability of BPH resistance in Rathu Heenati (RH), a traditional Sri Lankan rice cultivar containing qBph3, has made this strain one of the most popular hopper-resistance donors in the Mekong sub-region, where rice production is highly intensive. QTL mapping located qBph3 to the short arm of chromosome 6 between RM589 and RM588 based on KDML105 × Rathu Heenati (Jairin et al. [2009]) and PTB33 × RD6 (Jairin et al.[2007]) crossings. Other BPH resistance QTLs have also been reported, including Bph17 on chromosome 4 (Sun et al. [2005]), Bph4 on chromosome 6 (Kawaguchi et al. [2001]; Sun et al.[2006]) and Bph18 on chromosome 12 (Jena et al. [2006]). In addition to resistance QTLs, a putative sesquiterpene synthase (TPS) gene (Os04g27430) was identified using SFP mapping with isogenic lines derived from the backcross of RH and KDML105 (Kamolsukyunyong et al. [2013]). The TPS gene is induced after 5 days of BPH feeding. Functional markers were identified in exon 5 of the TPS gene that resulted in the deletion of seven amino acids in the susceptible rice line, as well as three additional SNPs associated with a transcriptional binding site, accounting for the differential response of TPS during the anti-feeding test (Kamolsukyunyong et al. [2013]).

Alkali disintegration has been used as a biomarker for gelatinization temperature (GT) in rice (Waters et al. [2006]; Kate-ngam et al. [2008]; Masouleh et al. [2012]). The alkali disintegration locus (ALK) was identified as the starch synthase IIa (SSIIa) gene on chromosome 6 that determines amylopectin structure (Bao et al. [2004]; Bao et al. [2006a]; Umemoto et al. [2002]; Umemoto et al. [2004]; Umemoto and Aoki, [2005]; Waters et al. [2006]; Lu et al. [2010]). Two functional SNPs were identified: GC/TT and G/A at positions 4329/4330 bp and 4198 bp, respectively (accession AY423717) (Bao et al. [2006b]; Umemoto et al. [2002]; Umemoto et al.[2004]; Umemoto and Aoki, [2005]; Waters et al. [2006]; Lu et al. [2010]). Successful marker assisted selection (MAS) programs utilizing the GT haplotypes for improving starch profiles have been reported (Tian et al. [2009]; Lu et al. [2010]).

'PinK3' is an aromatic, high-yielding, non-photoperiod-sensitive, high-amylose rice variety, but it is susceptible to BPH, BB, BL and Sub stresses. Here, we report the successful gene/QTL pyramiding of five functional genes (xa5, Xa21, Sub1A-C, SSIIa, TPS) and three QTLs (qBph3, qBL1, qBL11) into the 'PinK3' genome background using a multiplex pseudo-backcrossing approach based on MAS. The new, improved lines have a high-yield phenotype that confers submergence tolerance and resistance to BPH, BB and BL. This is the first report describing the application of pseudo-backcrossing to significantly shorten the time required for gene/QTL pyramiding in an annual crop (rice).

RESULTS

Streamline Gene Pyramiding in Rice

Four sets of donors (Table 1) containing a set of target genes/QTLs—Cholsub (Sub1A-C andSSIIa), Xa497 (xa5 and Xa21), Bph162 (qBph3 and TPS), and RBPiQ (qBL1 and qBL11)—were backcrossed in parallel once to 'PinK3' by targeting MAB to generate four sets of BC_1F_1 progeny. The resulting BC_1F_1 progeny were stepwise crossed to recombine four sets of the target genes/QTLs into a single set of pseudo-backcrossed progenies. By pair-wise crossing two pairs of BC_1F_1 lines, two sets of pseudo-backcrossed progeny—BC_2F_1 (Sub1A-C, SSIIa, xa5 and Xa21) and BC_2F_1 (qBph3, TPS, qBL1 and qBL11)—were generated (Figure 1). In cycle 4, all target genes/QTLs (Sub1A-C, SSIIa, xa5, Xa21, TPS, qBph3, qBL1 and qBL11) were recombined by crossing the two pseudo-backcrossed BC_2F_1 lines to generate 2,630 pseudo-backcrossed BC_3F_1progeny. Using target MAS and plant-type selection, 158 fully heterozygous pseudo-backcrossed BC_3F_1 lines were selected and selfed (cycle 5) to generate 11,405 F_2 progeny for large-scale, full-target MAS to generate 29 families for the target MAS purification. The numbers of positive plants (pseudo BC_3F_2) for all target genes/QTLs were segregated with Mendelian pattern (homozygous preference genotype = $1/4^n$). In cycle 6, selfing and full-target MAS yielded 29 best-selected, fully homozygous pseudo-backcrossed inbred lines (pseudo-BILs) carrying positive homozygous alleles of all of the donor genes, including Sub1A-C, SSIIa, xa5, Xa21, and TPS as well as qBph3 and qBL1-qBL11.

Table 1: The four donors and recurrent parents used in the multiple backcross gene pyramiding

Cultivar/ breeding line	Description	Variety type	Cross	Genotype on carrier chromosomes and target gene (s) or QTLs											References
				Sub	*xa5*	*Xa21*	qBL1	qBL11	Bph3	*TPS*	*Wx*	*SSi-ia*	*Os2AP*	*Hd1*	
PinK3	High yielding aromatic rice (pseudo-recurrent parent)	RIL	IR71501/ (KDMl105/ CT9993)	-	-	-	-	-	+	-	+	-	+	+	Rice Science Center (unpublished)
CholSub1	Submergence tolerance aromatic rice with high GT	RIL	IR57514/ KDMl105	+	-	-	-	-	-	-	-	+	+	+	Jantaboon et al. [2011]
Xa497	BLB resistance aromatic rice	BC_2F_2	IR62266/ KDMl105	-	+	+	-	-	-	-	-	-	+	+	Korinsak [2009]
RBPiQ	Blast resistance	RIL	JHN/ KDMl105	-	-	-	+	+	-	-	-	-	-	+	Rice Science Center (unpublished)
Bph162	Bph resistance photosensitive rice	BC_3F_6	Rathu Heenati/KDMl105	-	-	-	-	-	+	+	-	-	-	-	Jairin et al. [2009]

+ = desirable allele.

- = undesirable allele.

Ruengphayak et al.

Ruengphayak et al. Rice 2015 **8**:7 doi: 10.1186/s12284-014-0035-0

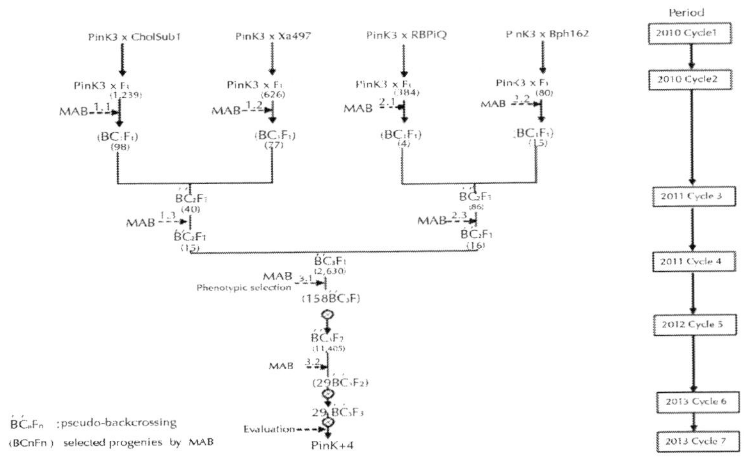

Figure 1: The gene/QTL pyramiding scheme used to generate the high-yield pseudo-BC BIL'PinK+4' line exhibiting submergence tolerance (Sub1A-C), bacterial leaf blight resistance (Xa21,xa5),rice blast resistance (qBL1andqBL11), brown planthopper resistance (qBph3andTPS) and desired cooking qualities (Wx, SSIIa, Os2AP).

Of these lines, five pseudo-BC_3F_3 BILs were chosen for field evaluation. The five pseudo BC_3F_3BILs (PinK+4) were selected based on completion of the target genomic regions with interesting starch profiles suitable for glycemic index research in the future (data not shown). In total, four donors and one recipient were intensively crossed and selected for seven cycles: two cycles to generate BC_1F_1, another two cycles to generate pseudo-backcrossed BC_3F_1, and three cycles of selfing to fix the final best-selected pseudo-BC_3F_3BIL for field evaluation.

Graphical Genotyping of Pseudo-BIL

To determine the effects of pseudo-backcrossing on genomic background composition, nine elite pseudo-BIL (BC_3F_2) families were selected for genome scanning using 61 SSRs, 35 and 26 of which were located on six carrier and six non-carrier chromosomes, respectively . The whole genome composition of the selected pseudo-BILs was characterized as the percentages of recurrent genome content (%RGC) and donor genome content (%DGC) based on the physical intervals of the SSR-based genome scanning using five BC_1F_2 (Table 2 and two BC_2F_2 (data not shown) and nine pseudo-BC_3F_3BILs (Table 3 and The %RGC and %DGC of BC_1F_2 ranged from 69.99 to 88.98% and from 11.02 to 30.01%, respectively whereas those variables for the pseudo-BC_3F_3BILs ranged from 74.50 to 81.30% and 18.70 to 25.50%, respectively The mean %RGC and %DGC values for BC_1F_2 were 80.04% and 19.96%, respectively (Table 2), whereas these values for the pseudo-BC_3F_3BILs were 77.48% and 22.52%, respectively (Table 3). These results indicated that at the BC_1F_1 step, there were no significant gains or losses of %RGC and %DGC from the two cycles of pseudo-backcrossing. For this reason, the theoretical RGC in pseudo-BC_3F_3 was not met.

Table 2: The percentage of genome composition (average per target locus) of the selected BC1F2lines resulting from backcrossing of four donors: CholSub1, Xa497, RBPiQ, and Bph162 on PinK3

% genome compositions (average per donor)									
	CholSub		Xa497		RBPiQ		qBph162	Ave.	Sum
	Sub1	**xa5**	**Xa21**	**qBL1**	**qBL11**	**TPS**			**qBph3**
Target gene/QTL	<0.01	<0.01	<0.01	0.03	0.26	<0.01	0.05	0.09	0.34
Donor segment link	1.81	0.00	3.77	1.59	0.00	0.00	0.00	1.79	7.17
Heterozygous segment link[1/]	0.00	0.83	0.13	0.12	0.95	0.00	1.16	0.80	3.19
Donor segment unlink	0.00	0.00	1.03	0.00	0.00	2.60	0.00	0.91	3.63
Heterozygous segment unlink[1/]	0.00	0.00	0.52	0.40	1.61	0.00	0.00	0.63	2.53
Sum of donor segments on target carrier chromosome	1.81	0.83	5.45	2.14	2.82	2.6	1.21	4.22	16.86
Donor segments on non-target carrier chromosome (4-5ch)								7.71	
[1/]Heterozygous segments on non-target carrier chromosome (4-5ch)								1.96	
Sum of donor segments on non-target carrier chromosome								13.89	
Donor segments on non-carrier chromosome (6 ch)								4.29	
[1/]Heterozygous segments on non-carrier chromosome								1.78	
Sum of donor segments on non-carrier chromosome								6.07	
Recurrent background								74.87	
%RGC								80.04	
%DGC								19.96	

[1/]= (1/2 total percentage of physical position).

Ruengphayak et al.

Ruengphayak et al. Rice 2015 **8**:7 doi:10.1186/s12284-014-0035-0

Table 3: The percentage genome compositions (average per target locus) in nine selected pseudo-BC3F3BILs

Region	% genome compositions (average per target)						
	Sub	***xa5***	**qBL1**	***Xa21*_qBL11**[1/]	***TPS***	**qBph3_*SSIIa***[2/]	**Sum**
Target gene/QTL	<0.01	<0.01	0.03	0.26	<0.01	0.05	0.34
Donor segment link	1.61	1.22	1.55	2.11	0.19	0.00	6.68
Heterozygous segment link	0.54	0.83	0.24	0.24	0.00	0.00	1.85
Donor segment unlink	0.00	0.18	1.39	1.40	1.44	0.47	4.88
Heterozygous segment unlink	0.00	0.00	0.04	0.23	0.00	0.76	1.03
Sum of donor segments on target carrier chromosome	2.15	2.23	3.25	4.24	1.63	1.28	14.78
Donor segments on non-carrier chromosome (6ch)							5.03
Heterozygous segments on non-carrier chromosome (6ch)							2.71
Sum of donor segments on non-carrier chromosome							7.74
Recurrent background							71.89
% RGC							**77.48**
% DGC							**22.52**

[1/]Xa21 located within qBL11 on chromosome 11.

[2/]SSIIa located within qBph3 on chromosome 6.

Ruengphayak et al.

Ruengphayak et al. Rice 2015 **8**:7 doi:10.1186/s12284-014-0035-0

We then looked into the distribution of donor genome segments across the six carrier and non-carrier chromosomes. For the carrier chromosomes, linkage drags were identified upstream and/or downstream of the donated target gene/QTL following transmission to the recipient genome. Even considering the stringent MAB on all target genes/QTLs during the BC_1 cycle, linkage drags were still detected for almost every donated locus, both homozygous and heterozygous, on one or both sides of the target gene/QTL, constituting more than half of the total donor segments on the carrier chromosomes (Table 2). The largest linkage drags were detected on both sides of the Xa21 locus (Table 2). However, only a small heterozygous linkage drag was detected at the xa5 locus. The contrast between the degree of linkage drag for the two functional genes Xa21 and xa5 was unexpected, as the pair of loci was inherited from the same donor, Xa497. As both xa5 and Xa21 were MAB using their functional markers, it is interesting to speculate on the differences in linkage drag between the two functional genes. The size of the linkage of Xa21 is much greater than for the two functional genes Sub1 and TPS, and the other QTLs, qBL1, qBL11 and qBph3. Heterozygous linkage drags were identified for five of the eight target genes/QTLs, including the Xa21 locus (Table 2), Donor-unlinks—the additional donor segments co-transmitted on the opposite (unlinked) chromosome arms of the target genes/QTLs—were identified on the carrier chromosomes containing Xa21, qBL1, qBL11 and TPS (Table 2). Therefore, the total donor segments transmitted via BC_1 along with Xa21 were obviously the largest among the target loci under MAB (Table 2).

Multiple target loci from donors are combined in the recipient genome, which include both linkage and non-linkage drags on the carrier chromosomes. Following two successive cycles of marker-assisted pseudo-backcrossing, BC_3F_3BILs linkage drags were detected innearly every case, with the exception of the qBph3-SSIIa locus. The total linkage drag (combined homozygous and heterozygous types) was 8.53%, which is more than half of the total donor component on carrier chromosomes (Table 3). The unlinkage drag values of the donated segments on the carrier chromosomes were between 5.91 and 40% of the total donor component on the carrier chromosomes (Table

3). To trace the potential origins of these large donor components, the sum of all donor compositions in the selected BC_1F_2 lines was compared with those components in the selected BC_3F_3BILs. On average, the donor genome content in BC_3F_3BILs was not significantly different from that of BC_1F_2 (Table 2 and Table 3). The same was true for the homozygous components of linkage drag and donor unlinked, whereas the heterozygous components were significantly decreased during pseudo-backcrossing. These results could indicate that high frequency of recombination between donated segments and the recipient genome contents of the carrier chromosomes in the BC_1 cycle was primarily due to the high donor genome content in successive pseudo-backcrossings. Furthermore, the significant reduction in heterozygous donor components for the BC_3F_3BILs highlights an advantage of pseudo-backcrossing in gene pyramiding.

For the non-carrier chromosomes, the total donor components on all six non-carrier chromosomes were not significantly affected (Table 2 and Table 3). However, the heterozygous donor components on both carrier and non-carrier chromosomes persisted, even in the selected pseudo-backcrossed BC_3F_3BILs. These results show that pseudo-backcrossing has similar effects as conventional backcrossing when considering the recovery of the genome content on non-carrier chromosomes.

Evaluation of the Pink+4 Phenotype

Field evaluations of agronomic and grain quality were performed based on complete target genes/QTLs, plant types, days to harvest and grain quality. The foreground selection successfully fixed homozygosity of the five target genes (Sub1A-C, SSIIa, xa5, Xa21 and TPS) and three QTLs for BL and BPH into approximately 77% of the genetic background of the pseudo-recurrent parent 'PinK3' (Table 3 and However, the majority of the advanced progeny exhibited significant phenotypic variation from their pseudo-recurrent parent 'PinK3' with respect to nearly all evaluated traits, with the exception of amylose content, grain length per width ratio and polished grain length (Tables 4 and 5).

Table 4: Trait evaluation and agronomic characteristics of four selected pseudo-BC3F3BILs (PinK+4) and parental lines with respect to submergence, bacterial blight, brown planthopper and blast resistance

No	Name	Family	Trait evaluation[1/]				Agronomic characteristics[2/]						
			Sub	BLB (TXO156)	Bph(UBN)	Blast (Mixed#2)	DM	NTP	**PH (cm)**	**NGP**	**PSF (%)**	**TGW (g)**	**GY (kg/ha)**
1	1E_06	1_H06	50.7$^{c3/}$	1.5^{a}	1.7^{a}	0.7^{a}	140.0^{e}	9.3^{ab}	70.1^{bc}	276.7^{e}	82.5^{ef}	42.9^{f}	7777^{ef}
2	20A09	4_E02	42.0^{bc}	1.4^{a}	1.7^{a}	0.7^{a}	140.0^{e}	9.3^{ab}	79.0^{c}	224.3^{d}	69.6^{c}	41.4^{e}	7381^{ef}
3	66B09	3_E03	31.3^{b}	0.8^{a}	1.7^{a}	0.6^{a}	128.0^{c}	9.2^{ab}	77.7^{c}	236.1^{d}	76.1^{d}	42.5^{f}	6977^{de}
4	78A03	13_H06	52.0^{c}	0.8^{a}	3.7^{abc}	0.0^{a}	143.0^{h}	9.2^{ab}	78.3^{c}	241.0^{d}	77.3^{d}	42.4^{f}	6966^{de}
5	117A08	2_A10	56.0^{c}	1.5^{a}	3.0^{ab}	0.0^{a}	126.0^{c}	8.6^{a}	77.3^{c}	168.7^{c}	81.1^{e}	41.5^{e}	6175^{cd}
6	PinK3	recurrent	0.0^{a}	12.8^{b}	6.3^{cd}	5.7^{d}	141.3^{f}	8.3^{a}	77.0^{c}	301.0^{f}	62.0^{b}	40.3^{d}	8306^{f}
7	Xa497	donor	3.3^{a}	0.5^{a}	6.3^{cd}	3.7^{c}	134.0^{d}	9.5^{ab}	59.3^{a}	166.0^{c}	84.9^{f}	40.5^{d}	5417^{c}
8	CholSub1	donor	50.7^{c}	nd	5.7^{bc}	2.3^{b}	122.0^{b}	10.7^{bc}	62.7^{ab}	120.3^{b}	75.0^{d}	39.4^{c}	4041^{b}
9	RBPiQ	donor	2.0^{a}	18.4^{b}	9.0^{d}	0.0^{a}	142.0^{g}	10.4^{cd}	74.3^{c}	185.7^{c}	94.1^{g}	34.8^{a}	5581^{c}
10	Bph162	donor	0.0^{a}	15.6^{b}	1.0^{a}	3.3^{c}	99.0^{a}	12.8^{d}	69.7^{bc}	81.0^{a}	58.5^{a}	38.5^{b}	2227^{a}

Remarks:

[1/]Trait evaluation.

Sub = Average % plant survival (%PS) after 15 days of flash flooding.

BLB = average lesion length in centimeters of the damage caused by the BB isolate TXO156.

Bph = Severity scores with UBN biotype at 9 DAI when TN1, the susceptible control died.

Blast = Average blast injury score when attacked by a Mixed#2 blast isolate from Thailand.

[2/]Agronomic characteristics.

DM (days to maturity), NTP (number of tillers per plant), PH (plant height from the soil surface to the neck of the panicle), NGP (number of grains per panicle), PSF (percent spikelet fertility), TGW (1,000 grain weight) and GY (grain yield).

[3/]Average values marked with different letters in the same column are significantly different at the 95% confidence level using LSD.

Ruengphayak et al.

Ruengphayak et al. Rice 2015 **8**:7 doi:10.1186/s12284-014-0035-0

Table 5: Grain quality and cooking quality traits of five pseudo-BC3F3BILs (PinK+4) as well as the donor and recurrent parents

No.	Name	BR[1/] (%)	HR[1/] (%)	GL/W[1/]	PRL[1/] (mm)	AC[1/](%)	ASV[1/]	CE[1/](%)
1	1E_06	72.9^{c2}	51.8^{c}	3.0^{ns}	0.75^{ns}	29.3^{c}	7.0^{c}	36.7^{abc}
2	20A09	72.9^{c}	52.7^{c}	3.1^{ns}	0.74^{ns}	27.6^{c}	7.0^{c}	43.3^{cdef}
3	66B09	71.5^{b}	52.1^{c}	3.1^{ns}	0.76^{ns}	27.5^{c}	2.0^{a}	31.5^{a}
4	78A03	71.2^{b}	51.3^{c}	3.2^{ns}	0.78^{ns}	27.1^{c}	2.0^{a}	31.2^{a}
5	117A08	73.0^{c}	53.1^{c}	2.9^{ns}	0.73^{ns}	29.5^{c}	7.0^{c}	46.7^{ef}
6	PinK3	72.4b^{c}	43.4^{b}	3.1^{ns}	0.74^{ns}	29.2^{c}	7.0^{c}	39.4^{bcd}
7	Xa497	71.9b^{c}	53.3^{c}	3.4^{ns}	0.74^{ns}	22.0^{b}	5.0^{b}	40.3^{bcde}
8	CholSub	73.2^{c}	42.4^{b}	3.4^{ns}	0.74^{ns}	14.5^{a}	2.0^{a}	44.4^{def}
9	RBPiQ	71.0^{b}	41.3^{b}	3.4^{ns}	0.68^{ns}	14.2^{a}	5.0^{b}	35.9^{ab}
10	Bph162	68.6^{a}	30.2^{a}	3.2^{ns}	0.70^{ns}	14.0^{a}	5.0^{b}	48.6^{f}

Remarks: [1/]BR (brown rice), HR (head rice), GL/W (grain length-width ratio), PRL (polished rice length), AC (amylose content), ASV (alkaline spreading value (1.7% KOH)), and CE (cooking elongation).

[2/]Average values marked with different letters in the same column are significantly different at the 95% confidence level using LSD.

Rice grains were harvested from the yield trial field during the wet season of 2012/2013.

Ruengphayak et al.

Ruengphayak et al. Rice 2015 **8**:7 doi:10.1186/s12284-014-0035-0

The uniformity of the pseudo-BC_3F_3BILs was the result of MAS. Even with respect to complex traits, such as grain yield, some of these progeny performed as well as the pseudo-recurrent parent 'PinK3' and significantly outperformed their donors (Table 4 and Figure 2). However, some progeny inherited inferior characteristics from the resistance donors, which affected maturity, grain numbers per panicle (NGP), % seed fertility (PSF) and grain yield (GY). In all cases, early-maturing progeny produced a lower grain number per panicle and lower grain yield (Table 4). Therefore, without rigorous background selection, pseudo-backcrossed progeny may not possess the desirable characteristics of their pseudo-recurrent parent.

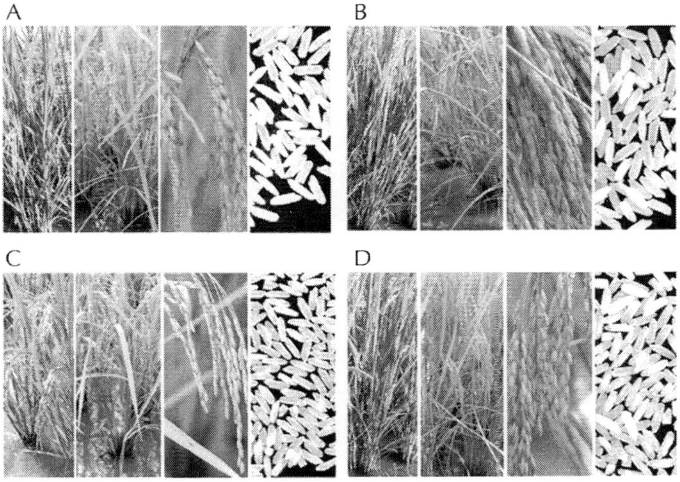

Figure 2: Plant and grain types of pseudo-BC BIL PinK+4 compared with PinK3 (pseudo-recurrent parent). A) PinK+4#1E06, B) PinK+4#20A09, C) PinK+4#66B09 and D) PinK3.

For the Sub1 selection, all progeny exhibited a significantly improved ability to withstand flash flooding during seedling stages compared with the susceptible 'PinK3' (Table 4). However, significant phenotypic variation (31–56%) was observed among the Sub1-Pink3 families compared with the control phenotype. Two reasons for this variation; First, the Sub1 gene was not directly inherited from the original FR13A but from CholSub1 donor one of the RILs from the mapping pop 'KD × FR13A' (IR57514) for submergence tolerance QTL by Siangliw et al. ([2003]). The second reason is the quantitative nature of multiple gene/QTL may partially regulated by recipient genetic background of CholSub1 donor and the 'PinK3', the highly susceptible recipient. Furthermore, the fact that significant variations in traits related to submergence tolerance were observed among the individual progeny of IR57514/Kao Dawk Mali 105 (Jantaboon et al. [2011]) is also consistent with our findings, indicating the quantitative nature of such traits.

For BB, selection based on xa5 and Xa21 resulted in significant improvements in bacterial leaf blight resistance, specifically to the TXO156 virulent isolate (Table 4) and to several BB isolates identified in Thailand For leaf/neck blast resistance, QTL pyramiding of the two QTLs on chromosomes 1 and 11 onto the 'PinK3' background successfully improved resistance to a wide range of blast isolates collected in Thailand (Table 4).

For BPH resistance, based on the UBN biotype (a BPH biotype that has been well characterized in Thailand), the progeny exhibited significant improvement over their pseudo-recurrent parent 'PinK3' but were not as resistant as the resistance donor. The Bph3 QTL and the TPS gene from the Bph162 donor were co-inherited from the broad spectrum BPH resistance cultivar 'Rathu Heenati' in crosses with KDML105, as reported by Jairin et al. ([2009]). Some selected introgression lines from that report were moderately resistant (MR) to various BPH biotypes. It appears that introgression of only qBph3 and TPS from Bph162 in pseudo-backcrossed BILs was not enough to withstand some of the BPH biotypes used in our experiments.

All selected pseudo-BC_3F_3 BILs (PinK+4) contained the aromatic allele and the Wxa allele known to confer high amylose content and a starch profile suitable for further analysis and industrial food applications.

DISCUSSION

Streamline Backcrossing Design

In a gene-pyramiding project, several donors, each providing a target QTL with flanking markers, are used as parental sources for new traits with the goal of improving a favorable variety that has a preferred genetic background. To best facilitate the efficient integration of multiple QTLs into a single optimal variety, an improved breeding platform was developed based on pseudo-backcrossing. In conventional backcross breeding experiments, nearly isogenic lines from each donor were developed prior to pyramiding (Luo and Yin, [2013]; Singh et al. [2013]) to recover the background genome of the recurrent parent. However, when introducing multiple traits, this approach can be tedious and time-consuming. This novel platform based on pseudo-backcross breeding involves both multiple foreground selection and background genome recovery in an abridged manner. Within seven cycles—consisting of a single backcross, two cycles of pseudo-backcrossing and three cycles of line fixation—the entire project was accomplished within four years.

Recurrent Genome Background Recovery

Transferring multiple resistance genes using conventional MAB requires at least three to four backcrosses to guarantee a high recovery of the recurrent parent phenotype (Joshi and Nayak, [2010]; Suh et al. [2013]). In this study, without background selection, the pseudo-BC_3F_3BILs (PinK+4) contained 74.50 to 81.30%RGC, which is significantly below the theoretical value of 93.75% possible following three conventional backcrossings. When comparing graphical genotyping among BC_1F_2 progeny and pseudo-BC_3F_3BILs, the total %DGC was only slightly increased. These results indicate that without background selection, pseudo-backcrossing can only maintain the %RGC gained during the first backcross generation. A low background recovery rate was also reported for the introgression of stripe rust resistance in wheat. Without marker-assisted background selection, the %RGC was only 82% in BC_4F_7 progeny (Randhawa et al. [2009]). However, when combined with phenotypic selection, %RGC was improved to 85–92% in BC_3(Sundaram et al. [2008]; Korinsak et al. [2011]; Singh et al. [2013]). Indeed, the utilization of genome-wide molecular markers for background screening during backcrossing has been suggested as the best method for improving low %RGC (Rajpurohit et al. [2011]; Suh et al.[2013]).

The number of molecular markers used for genome-wide scanning, which ultimately determines cost vs. precision, has varied from 44 to 205 SSR loci in rice. Four groups have reported rice MAB projects involving background selection: Group 1) used 44–51 SSRs (Yi et al. [2009]; Siangliw et al. [2003]; Tomita [2009]; Wongsaprom et al. [2010]), Group 2 used 67–72 SSRs (Win et al.[2012]; Singh et al. [2013]), Group 3 used 84–97 SSRs (Siangliw et al. [2007]; Jantaboon et al.[2011]), and Group 4 used 107–205 SSRs (Rajpurohit et al. [2011]; Korinsak et al. [2011]; Suh et al. [2013]). However, the majority of these MAB projects utilized molecular markers during the final stage of selection.

Linkage Drag

In conventional backcrossing, many portions of the donor genomes are inserted into both the carrier and non-carrier recipient chromosomes during early cycles. After continued backcrossing, the donor genome segments are gradually replaced by sequences from the recurrent parent at varying rates. In most backcrossing programs, linkage drag is responsible for long-lasting donor genome segments remaining in the recurrent genome. However, there is no difference in terms of linkage drag in pseudo-backcrossing schemes. In pseudo-backcrossing BILs, there was only a 1% reduction in linkage drag from BC_1 to pseudo-BC_3, and most of this reduction was due to the heterozygous segments of the linkage drag. Our results also revealed that the degree of linkage drag is independent of the size of the target genes/QTLs. When comparisons were made between the selection of single genes or single QTLs, the degree of linkage drag was less for QTLs than for single genes. Of the five single gene selections, SSIIa, Sub1, xa5, Xa21 and TPS, only selection for Xa21 showed large, persistent linkage drag. Persistent linkage drags when selecting for BB genes have been reported in several backcross breeding programs, such as during the transfer ofXa4 + xa5 + Xa21 from indica 'IRBB57' into japonica 'Mangeumbyeo' (Suh et al. [2013]), and pyramiding of the BB resistance genes Xa21 and xa13 and a semi-dwarfing gene (sd-1) from PR106-P2 into Type 3 Basmati (Rajpurohit et al. [2011]). The problem of persistent linkage drag when selecting for Xa21 may due to the fact that Xa21 was derived from IRBB21, which inherited the chromosomal region containing Xa21 from the wild species Oryza longistaminata through several cycles of backcrossing with indica rice (Song et al. [1995]). The degree of linkage drag may depend on linkage disequilibrium surrounding the target gene to be transferred. Genes inherited from wild species in cultivated strains may be retained within long, stable LD stretches that are difficult recombine. Therefore, selection based on functional markers alone does not guarantee linkage drag-free progeny. More successful single-gene target selection has been reported when markers flanking the gene of interest were also selected for (Rajpurohit et al.[2011]). In one of the most comprehensive backcrossing projects, a single gene, sd1, was integrated into the desirable variety 'Koshihikari' using eight cycles of MAB with 51 SSR markers surrounding sd1 to completely eliminate linkage drag (Tomita [2009]).

Grain Yield Performance of Pseudo-BC BILs

In this study, we combined multiple resistance genes from four donors using a new backcrossing method involving pseudo-backcrossing. The results show that all pseudo-BC BILs showed significant improvements in resistance to BB, BL, BPH and Sub compared with the pseudo-recurrent parent 'PinK3', as well as significant improvements in grain yield (21–68% over the donors, but 7–26% lower than the recipient). The reduction in grain yield in the pseudo-BC BILs should be interpreted in several ways. First, there was an average of 7.6% and 22.5% of linkage drag and DGC, respectively, in the recurrent genome. As these donors were inferior in grain yield with respect to the pseudo-recurrent parent 'PinK3', the high persistent %DGC could have disrupted the optimal expression of high-yield genes in pseudo-BC BILs. Second, there can be a slow recovery of the %RGC during pseudo-backcrossing when MAS for the recurrent background is not in place. Under such conditions, reconstruction of the recurrent genome content by recombining different pseudo-backcrossed lines is not favorable, as different donor segments on both carrier and non-carrier chromosomes have more chances to recombine, creating new substitution lines that do not resemble the recurrent parent. Third, the multiple donated genes/QTLs from donors to the recipient act as a 'genetic load' against the fitness of the recurrent parent. The over-expression of multiple resistance genes could counteract the metabolic energy needs necessary for high yield. Therefore, pseudo-backcrossing may be the fastest method for gene/QTL pyramiding, although it may not be the ideal breeding platform for creating elite recurrent varieties. However, marker-assisted, genome-wide scanning can be implemented during early stages to facilitate the reconstruction of favorable genomic backgrounds at the end of the pseudo-backcrossing scheme. Ultimately, the trade-offs must be considered by the breeders. Ideally, new high-throughput, low-cost, genome-wide scanning technologies should be utilized in combination with skillful breeder selection. For the whole project, more than 50,000 plants were individually genotyped for at least one molecular marker. If budget is allow, extensive background selection must be emphasized.

CONCLUSION

We improved high-yield, aromatic rice varieties by introducing desirable multiple traits by pyramiding five target genes (Sub1A-C, SSIIa, xa5, Xa21 and TPS) and three QTLs (qBph3, qBL1 and qBL11) from four resistance donors. We redesigned the gene-pyramiding platform to minimize the total project time span by integrating MAS into pseudo-backcross breeding. Consequently, only seven breeding cycles in four years were required to develop new varieties exhibiting multiple resistance traits. Using pseudo-backcrossing, approximately 77.48% of the recurrent genome background was recovered. With additional background genome selection, the recurrent genome background can further improve the %RGC and optimize the expression of introgressed QTLs.

METHODS

Plant Materials Used

PinK3 is a high-yield, irrigated aromatic rice cultivar developed by Rice Science Center, Kasetsart University, Thailand (unpublished). However, this variety is susceptible to flash flooding (Sub), bacterial leaf blight (BB), leaf-neck blast (BL) and the brown planthopper (BPH). The four donors used to transfer five genes and three QTLs to the pseudo-recurrent parent 'PinK3' (Table 1) were developed by the Rice Gene Discovery and Rice Science Center, Kasetsart University, Thailand. The four donors used to improve the abiotic and biotic stress tolerances of 'PinK3' are listed in Table 1.

Pseudo-Backcross Design

The pseudo-backcross platform is divided into three steps. In the initial step, one round of backcrossing is conducted to donate the favorable QTL allele to the recipient background using marker-assisted backcrossing (MAB). Each QTL-BC_1F_1 contains approximately 75% of its recurrent genome content (RGC). In the second step, the BC_1F_1-plus QTLs are used as pseudo-recurrent parents, and pseudo-BC_2F_1 plants are formulated by crossing between them. More BC_1F_1 plus new QTLs can be crossed to successively generate pseudo-BC_nF_1 QTLs and, thus, to continue streamline gene pyramiding. At the end of this step, the BC_nF_1, which contains the new QTLs at full heterozygosity at all target marker loci, are self-pollinated to fix the target loci In rice, this new platform allows breeders to stack more QTLs in the shortest possible time (shorter than that required by any other method).

Donors for Gene Pyramiding in Rice

Four donors providing submergence tolerance, bacterial leaf blight resistance, blast resistance, BPH resistance and desired gelatinization temperature were introduced into PinK3 (aroaro andWx^AWx^A) as the female pseudo-recurrent parent. The donor CholSub1 carries two target traits: submergence tolerance and desired gelatinization temperature (Sub1A-C and SSIIa); the donor Xa497 carries two functional genes for bacterial leaf blight resistance (xa5 and Xa21); the donor RBPiQ carries two QTLs for blast resistance (qBL1 and qBL11); and the donor Bph162 carries two target traits for BPH resistance (qBph3 and TPS) (Table 1). These four donors were used for gene pyramiding.

Genomic DNA Isolation

Rice seedlings from each segregating population were grown in a 288-well plastic tray (representing three 96-well plates). Young leaves from 14-day-old individual plants were cut into small pieces and placed (~0.2 g weight per sample) in a 2-ml 96-well plastic block. Leaf tissues were ground in liquid N_2 using a Tissue Striker II (KisanBio, Seoul, South Korea). After grinding, 300 μl Agencourt®Chloropure lysis buffer was added to the samples. Homogenized tissues were incubated in a 2-ml 96-well plastic block at 65°C for 1 hour. The sample blocks were then centrifuged at 4,000 rpm for 10 minutes. Lysates (containing at least 200 μl) were transferred to a new 2-ml 96-well plastic block using an Automated Biomeck NX AP96 instrument (Beckman Coulter, California, USA). The extraction was conducted using the standard protocol of Agencourt Chloropure for nucleic acid isolation from plants (Beckman Coulter, California, USA).

Foreground MAS

Foreground selection was performed using two marker systems. For SNP and functional markers, multiplex genotyping was conducted using the SNPstream system (Beckman Coulter, California, USA). The remainder of the foreground markers were SSR markers flanking specific QTLs. The SNP-based genotyping array was described.

High-Throughput Multiplex SNP Genotyping

High-throughput genotyping was performed by multiplex PCR, as described by Bell et al. ([2002]) with certain modifications. In brief, the forward/reverse (18–20 nt in length) and SNP-specific (40–45 nt in length) primers were designed for each foreground locus (Primers can be manually designed) to generate a product of 90–180 nt in size. The program selects the best Single Base Extension (SBE)-primer based on sequence melting temperature (T_m; °C) and secondary structure. At the 5′ end of the SBE-primer sequences are 20-nt tags that are complementary to the sequences of specific positional tags in the SNPware (384-well) microarray format (Beckman Coulter, California, USA) (Table 6).

Table 6: Primer sequences used in the SNP genotyping format for foreground selection

Multiplex PCR

No.	SNP name	Chr.	Trait	Tag number	SNP type	Amplicon size (bp)	Forward primer (5′---3′)	Reverse primer (5′---3′)	SNP-specific primer (5′---3′)
1	Sub1C	9	Sub	9	A/G	111	ACGAGCCGAC-GACGACGA	ATCTCCGACG-CCCACCTC	CCGCCAGTAAGACCT-AGACGCGGCGGCGG-CGGCGGCGGAGGGAGA
2	xa5*	5	BB	2	C/G	303	GGCCACCTT-CGAGCTCTACC	CAACATTGCAA=CT-CCGTGATAAG	CTCAGACTACGAAT-CCACGTGTAAAGTAGAT-ACCTTATCAAACTG
3	Xa21	11	BLQ11	40	T/C	141	AAAGCTAG-GCTGCTAGTGCTG	AAAATAGTATATA-TGTACCACTGCTTCTT	AAGTACCACGTCAA-CGTCACTATGCTTCAAGG-TCAGGGTGGTCGA
4	SSIIa	6	GT	25	G/T	89	CCACTGCCT-CGAGACGTA	CGTGGTCCCA-GCTGAGGT	CCATAACAACTTA-CCAGCCAGCAAGTAC-AAGGAGAGCTGGAGGGG
5	Waxy	6	AC	31	G/T	141	TTCACTTCTC-TGCTTGTGTTGT	TACTTGTAAG-GAAAAACGAGCAA	CAGAACATCCTCAG-AAGCAAGTTCATCAG-GAAGAACATCTGCAAG
6	*aromarker**	8	FR	41	A/G	400	AATCATGTATA-CCCCATCAA	TTTCCACCA-AGTTCCAGTGA	TACCTATGACCAG-CAAGCACAACCTTAA-CCATAGGAGCAGCTGAA
7	Hd1	6	Photo	13	C/G	152	TCCAAAGATT-CCGACAACA	TTGTCGTAGTA-CGAATTGTACCC	CAACAATACGAGCC-AGCAAGACAACAACA-ACGACAACGACAATAA
8	TPS_Chr4_ ATHB1*	4	Bph3	27	T/G	264	AAGCGCTTA-TATTCAAGCAGAA	TCCATTCTT-CCGATCTCTGG	GCAAGCCATCAGCT-AATACATTCATGAAACA-GTTCTAGCAATAAT
9	Bph_ Chr6_1210*	6	Bph3	17	T/C	287	GAAAGCCTT-TGAAACAAAGTATTGA	CTTGAATTTGA-AGTTGATTTTAGGG	GCAGACAACGAACAA-CTACCAAACGGCATAT-TTGCAAACAGAAAA

10	Bph_Chr6_3380*	6	Bph3	22	T/C	300	AGAGGAAAT-GATTCAAGGAG	AGCTAGCAG-GCGTAGCTTAT	GCAACATAAGACCGC-TCAACTAGTTAATTT-CACGCCATGACAGAT
11	Bph_Chr6_1380*	6	Bph3	19	A/G	296	TTTTGTTTTC-TTCTTGAGAGTGGTC	CAAGGTAATG-ACATCAAGAACCAAT	CCACTCAACTCCACGA-ATACCACATGTTTA-TTTTTAATTTCACAG
12	TBGI055716	1	BLQ1	24	T/A	169	ACGATGCG-GCACTCGTCG	TGTTCTTGAAC-GCGGCGA	CAACAAGACATA-ACAACGCAGTGGAGTGG-TGGATGAAGCGGAAGA
13	TBGI055578	1	BLQ1	42	A/G	93	ATTTGCTGC-TCATGGTGG	TGGGGAAGCC-GAGGAGAT	CTCACTATCTGACAAG-CCACGTGGAGCCTC-CTCACCAGGAAGTGC
14	TBGI055841	1	BLQ1	46	T/C	115	ATTGGCATC-GTTTGGTCTG	ATTCCGTGCATAT-ATACGAACTTC	ACAGATCACTCACCG-ACTAAATACCTGCGTCG-AGTAGAGACGATG
15	TBGI454069	11	BLQ11	14	A/G	142	TGAATTGTC-GTCCTCTAACAAC	TATGGAATATCT-GCATCATGAAC	AACATACAGACGCACT-CCTCTCTTTGCTATGCC-AAAGTCTGCTAC
16	TBGI453598	11	BLQ11	23	C/T	90	TTTGCTGTG-ACGGGAAGA	AAAAAGGAAC-TAGCCAGTTTTG	AGTAGCCTAACAGCACT-CGAAAGATCGAGTGC-TCTATTGCAACCG
17	TBGI453126	11	BLQ11	32	C/T	92	ACCGACGC-TGCTGCAGAA	AGCGGTTATG-GATGGCTAA	CAAGCAACGACCTACTAC-AAGCATGGCGTTTGA-GCGCGTCCTGGG
18	TBGI454717	11	BLQ11	36	T/A	132	ATCCTACC-GTCCGCTCTG	AATTCGGTCTT-CGTAAACACG	CACCGCTATCAACAGA-CTTGGTGCGGTAGTTT-CTGGGAAGCTACG
19	TBGI454800	11	BLQ11	38	G/C	113	TACTACAAC-AACAGGAACGCC	TTGATGATGAA-GTGGATGAGC	ACGTAAGACCACTCAAGA-CCAGAAGACGCTGAAC-AGGATGGCGAT

*These primer designs were optimized manually.

Ruengphayak et al.

Ruengphayak et al. Rice 2015 **8**:7 doi:10.1186/s12284-014-0035-0

Multiplex PCR

A 10-µl PCR reaction containing 5 µl genomic DNA (10 ng/µl) and 5 µl KAPA TaqHotStart PCR buffer (1 U KAPA TaqHotStart (KapaBiosystems, MA, USA) (final concentration of 1× KAPA TaqHotStart Buffer: 75 µM dNTPs, 5 mM $MgCl_2$ and 50 nM 38-primer pool) was performed in a 384-well PCR plate (Sorenson BioScience, UT, USA). The following thermocycler touch-up PCR cycle was used: 95°C for 3 min, followed by 6 cycles of 95°C for 30 seconds, 50°C up to 55°C (0.3 increment/cycle) for 30 seconds, and 72°C for 30 seconds; this was followed by 34 cycles of 95°C for 30 seconds, 55°C for 30 seconds, 72°C for 30 seconds and a final extension at 72°C for 7 min. Subsequently, the temperature was held at 4°C.

Multiplex PCR assays were prepared separately based on the SNP panel type (A/G, A/C, A/T, G/C, G/T and C/T). Following PCR amplification, PCR products were cleaned, and the SBE reactions were performed. Next, multiplex SBE products from different panel types were pooled prior to the hybridization step.

PCR clean up, SBE reactions, hybridization and washing, SNPstream imaging and data analysis were performed as described by Bell et al. ([2002]).

Evaluation Of Abiotic and Biotic Stress Traits

Submergence Screening

The parents and pseudo-BC_3F_3BILs (PinK+4) were screened for submergence resistance traits. The experiment was conducted under complete submergence in an outdoor lagoon located at the Rice Science Center, Kasetsart University, Kamphaeng Sean Campus, Thailand, during the dry season of 2013. The experiment was arranged using a randomized complete block design with three replications. Sixteen three-week-old BC_3F_3BIL seedlings and controls including PinK3, CholSub, Xa497, RBPiQ and Bph162 were transplanted in three replicate plots (plot size: 0.75 × 0.75 m^2) at a spacing of 25 cm × 25 cm. Two weeks after transplanting, the number of seedlings was counted in each plot; then, the lagoon was filled with water to a depth of 2 m. To impose severe submergence stress, the seedlings were completely submerged for 15 days; the water level was maintained at

1–1.2 m above the leaf tip throughout the experimental period. After this period, the lagoon was drained, and the seedlings were re-exposed to air for 10 days (Jantaboon et al. [2011]). The number of surviving plants was recorded. The percentage of survival (PS) was calculated using the following equation:

$$PS = \frac{\text{Number of surviving plants}}{\text{Total number of plants}} \times 100$$

Brown Planthopper (BPH) Screening

A set of pseudo-BC_3F_3BILs (PinK+4) and their parents were screened for resistance against BPH using standard seedbox screening (SSBS); the BPH population used was collected from Ubon Ratchathani provinces (Jairin et al. [2007]). The SSBS was conducted at the seedling stage (10 days old) under greenhouse conditions following the method described by Heinrichs et al. ([1985]). Damage scores were recorded when the susceptible control, 'TN1', died (9 days after infestation; 9DAI), using the standard evaluation system (IRRI [1996]).

Bacterial Leaf Blight Screening

The Xoo isolate TXO156 was selected for this experiment. The isolate was grown following the method described by Win et al. ([2012]). The parents and pseudo-BC_3F_3BILs (PinK+4) were grown in a greenhouse for 30 days before inoculation. The inoculation procedures used were adapted from those described by Korinsak et al. ([2009a], [b]). Three to four fully expanded leaves of each plant were inoculated. Lesion length (LL) was measured at 12–14 days after inoculation. The resistance reaction was classified as resistant (R), moderately resistant (MR), moderately susceptible (MS) and susceptible (S) when the values of LL were 0–3 cm, 3.1–6 cm, 6.1–9 cm and more than 9 cm, respectively (Yang et al. [2003]; Lin et al. [1996]).

Leaf Blast Screening

Thailand Magnaporthe oryzae mixed isolates#2, including THL710 (Mae Hong Son), THL282 (Phrae), THL906 (Yala), THL122 (Chiang Rai), THL757 (Mae Hong Son) and THL603 (Surin) (Rice Gene Discovery,

Thailand, unpublished), which can damage the PinK3 form in the mixed isolate pre-screening, was used in leaf blast screening experiments. The inoculum was prepared and the plants were inoculated following the method described by Marchetti et al. ([1987]) with some modifications. Pseudo-BC_3F_3BILs (PinK+4) and their parents were grown in polyvinyl trays containing paddy field soil (four seedlings per line) following a three-replication completely randomized design (CRD). The seedlings were maintained in a greenhouse for 17 days before inoculation, after which they were inoculated with mixed isolates#2. Disease scoring was recorded at seven days after inoculation on a 0 to 6 scale following the procedure described by Roumen et al. ([1997]) and IRRI ([2002]). The average score of each line was computed from the disease score measured for 12 individual plants.

Recording of Important Agronomic Traits

Traits measured included days to 100% flowering (DF100), days to maturity (DM), number of tillers per plant (NTP), plant height (PH), number of grains per panicle (NGP), percent spikelet fertility (PSF), 1,000 grain weight (TGW) and grain yield (GY); these traits were measured for rice plants grown during field trials at Kasetsart University, Kamphaeng Sean, Nakhon Pathom, Thailand.

Twenty-one-day-old seedlings were transplanted in three replicates in 1 × 2 m^2 plots using 25 × 25 cm^2 plant spacing. Agronomic traits were recorded for five randomly selected plants grown in each plot. The DF was recorded when 100% of the individual plants in each plot flowered. NTP, PH, NGP and DM were measured at maturity, and the results were averaged from five randomly selected plants in each plot. PH was measured from the soil surface to the neck of the panicle. The NGP was counted manually for five panicles. To measure the GY in each plot, only the inner rows (containing 21 plants) were used. Two border rows on each side of the plot and the border plants of each row were discarded. The GY recorded for each plot was adjusted to 14% moisture content and then extrapolated to units of kg per ha. TGW measurements were replicated three times. Statistical analysis was performed using the STATGRAPHICS plus 3.0 software package (Manugistics [1997]).

Evaluation of Grain Quality

Grain quality was evaluated using grain harvested from the trials field. Rice grains of the pseudo-BC_3F_3BILs (PinK+4) and their parents were harvested at physiological maturity and sun-dried in a greenhouse. The dried grains were stored at room temperature for one month prior to the grain quality traits evaluation. Three hundred grams of grains was sampled from each replicate. The grains were mechanically dehulled and polished using a mini-polisher. Four physical grain qualities, including percentages of brown rice (BR), head rice (HR), grain length (GL/W) and %cooking elongation increased (CE), were evaluated for the polished rice. Ten grains of paddy rice were measured using a Vernier caliper, and the GL/W ratio was calculated. The polished rice grain length (PRL) was measured using the same method. The cooking elongation of the polished rice was determined by boiling 20 grains in 5 ml of dH_2O for ten minutes. Cooked grain lengths were measured after air-drying the grains for 1 hour. Two chemical grain qualities, amylose content (AC) and gel temperature (GT), were evaluated following the procedures described by Lanceras et al. ([2000]). GT is an indicator of the time required for cooking. The GT was indirectly estimated based on the alkali spreading value (ASV); higher values of ASV represent increased spreading in alkali and therefore represent lower values of GT; conversely, smaller values of ASV indicate higher values of GT.

Genomic Scanning of Foreground and Background

Using simple sequence repeat (SSR) markers spanned across the genome, the effects of foreground selection on linkage drag and genome background content were estimated using the graphical genotyping software GGT 2.0 (Berloo [2007]) based on a specific F_2 population type calculation. The effects of marker-assisted selection were studied on BC_1F_2 and pseudo-backcross progeny after recombining all target genes/QTLs (PinK+4). Five selected BC_1F_2 and nine pseudo-BC_3F_3BILs (PinK+4) representing nine recombined families were analyzed for genetic background recovery; 61 SSR markers showing clear polymorphism between parents were used Twenty-six markers distributed over six non-carrier chromosomes (chromosomes 2, 3, 7, 8, 10 and 12), as well as 35 SSR markers distributed over six carrier chromosomes (chromosomes 1, 4, 5, 6, 9 and 11) were used for background scanning. Furthermore, QTL-specific markers located within each QTL were developed to estimate the risk of target loss. Definitions of parameters describing % recurrent genome content (%RGC) and % donor genome content (%DGC) were calculated according to Xi et al. ([2006]), (Suh et al. ([2009]) and Suh et al. ([2013]) with some modification.

Donor segment link: homozygous allele similar to each donor on the same arm of the carrier chromosome.

Heterozygous segment link: heterozygous allele on the same arm of the carrier chromosome.

Donor segment unlink: homozygous allele similar to each donor on the other arm of the carrier chromosome.

Heterozygous segment unlink: heterozygous allele on the other arm of the carrier chromosome.

Donor segments on non-carrier chromosome: homozygous allele similar to each donor on non-carrier chromosomes.

Heterozygous segments on non-carrier chromosomes: heterozygous allele on non-carrier chromosomes.

Recurrent background: homozygous allele similar to the pseudo-recurrent 'PinK3' allele.

AUTHORS' CONTRIBUTIONS

SR and SP performed the cross-hybridization and MAB screening experiments. EC developed the SNP marker set used for MAB. WK and WS performed the BPH infestation experiments and identified candidate genes for BPH resistance. EP performed the submergence screening. SK and SK performed the BB and BL screenings. The entire study was designed and coordinated by AV. SR drafted the manuscript, and AV corrected the manuscript's draft. All authors read and approved the final version of the manuscript.

ACKNOWLEDGMENTS

This work was supported by the Agriculture Research Development Agency (ARDA) (Grant No. P12/2552). SR gratefully acknowledges financial support from the Royal Golden Jubilee (RGJ)-PhD program, Grant No. PHD/0009/2546, from the Thailand Research Fund (TRF).

REFERENCES

1. Ashikari M, Matsuoka M: Identification, isolation and pyramiding of quantitative trait loci for rice breeding. Trends Plant Sci 2006, 11(7):344-350.
2. Bao JS, Sun M, Zhu LH, Corke H: Analysis of quantitative trait locus for some starch properties in rice (Oryza sativa L.), thermal properties, gel texture, swelling volume.
3. J Cereal Sci 2004, 39:379-385.
4. Bao JS, Corke H, Sun M: Nucleotide diversity in starch synthase IIa and validation of single nucleotide polymorphisms in relation to starch gelatinization temperature and other physicochemical properties in rice (Oryza sativa L.). Theo App Genet 2006, 113:1171-1183.
5. Bao JS, Shen SQ, Sun M, Corke H: Analysis of genotypic diversity in the starch physicochemical properties of non-waxy rice: apparent amylose content, pasting viscosity and gel texture. Starch-Starke 2006, 58:259-267.
6. Bell PA, Chaturvedi S, Gelfand CA, Huang CY, Kochersperger M, Kopla R, Modica F, Pohl M, Varde S, Zhao R, Zhao X, Boyce-Jacino MT: SNPstream UHT: ultra-high throughput SNP genotyping for pharmacogenomics and drug discovery. Biotechniques 2002, 32:S70-S77.
7. Berloo RV: GGT 2.0 user manual. Laboratory of plant breeding. Wageningen University, Netherland; 2007.
8. Bouman BAM, Lampayan RM, Tuong TP (2007) Water management in irrigated rice: coping with water scarcity. International Rice Research Institute (IRRI), Los Baños,Philippines. http://www.knowledgebank.irri.org/ewatermgt/courses/course1/resources/books/water_management_1.pdf.

9. Bouquet A: Introduction dansl'espece Vitis vinifera L. d'un caractere de resistance al'oidium (UncinulanecatorSchw. Burr) issul'espece Muscadinia rotundifolia (Michx) Small. Vignevini 1986, 12(suppl):141-146.
10. Heinrichs EA (1979) Chemical control of the brown planthopper. In: Brown planthopper: Threat to rice production in Asia, IRRI Publications, Los Baños, Philippines, pp. 145-167.
11. Heinrichs EA, Medrano FG, Rapusas HR: Genetic Evaluation for Insect Resistance in Rice. IRRI, Los Ban˜os, Philippines; 1985.
12. IRRI: Standard Evaluation System for Rice. IRRI, Manila, Philippines; 1996.
13. IRRI: Standard Evaluation System for Rice (SES). International Rice Research Institute, Los Banos, Philippine; 2002.
14. Jairin J, Phengrat K, Teangdeerith S, Vanavichit A, Toojinda T: Mapping of a broad-spectrum brown planthopper resistance gene, Bph3, on rice chromosome 6.
15. Mol Breed 2007, 19:35-44.
16. Jairin J, Teangdeerith S, Leelagud P, Kothcharerk J, Sansen K, Yi M, Vanavichit A, Toojinda T: Development of rice introgression lines with brown planthopper resistance and KDML105 grain quality characteristics through marker-assisted selection. Field Crop Res 2009, 110(3):263-271.
17. Jantaboon J, Siangliw M, Im-mark S, Jamboonsri W, Vanavichit A, Toojinda T: Ideotype breeding for submergence tolerance and cooking quality by marker-assisted selection in rice. Field Crop Res 2011, 123(3):206-213.
18. Jena KK, Jeung JU, Lee JH, Choi HC, Brar DS: High-resolution mapping of a new brown planthopper (BPH) resistance gene, Bph18(t), and marker-assisted selection for BPH resistance in rice (Oryza sativa L.). Theor Appl Genet 2006, 112:288-297.
19. Joshi RK, Nayak S: Gene pyramiding-a broad spectrum technique for developing durable stress resistance in crops. Biotechnology and Molecular Biology 2010, 5(3):51-60.
20. Kamolsukyunyong W, Sukhaket W, Ruanjaichon V, Toojinda T, Vanavichit A: Single-feature polymorphism mapping of isogenic rice lines identifies the influence of terpene synthase on brown planthopper feeding preferences. Rice 2013, 6:18.

21. Kate-ngam S, Kimchaiyong W, Wanchana S, Toojinda T: Association analysis and functional marker development of soluble starch synthase IIa (SSIIa) and gelatinization properties in Thai rice. In Proceeding of the 5th International Crop Science Congress & Exhibition: Recognizing Past Achievements, Meeting Future Needs April 13–18, 2008. International Convention Center, Jeju, Korea; 2008.
22. Kawaguchi M, Murata K, Ishii T, Takumi S, Mori N, Nakamura C: Assignment of a brownplanthopper (NilaparvatalugensStål) resistance gene bph4 to the rice chromosome 6. Breed Sci 2001, 51:13-18.
23. Korinsak S. Marker-assisted pyramiding bacterial blight resistance genes (xa5, Xa21, xa33(t), Xa34(t) and qBB11) in rice. Dissertation: Kasetsart University; 2009.
24. Korinsak S, Sirithanya P, Toojinda T: Identification of SSR markers linked to a bacterial blight resistance gene in rice cultivar 'Pin Kaset'. KKU Res J (GS) 2009, 9:16-21.
25. Korinsak S, Sriprakhon S, Sirithanya P, Jairin J, Korinsak S, Vanavichit A, Toojinda T:Identification of microsatellite markers (SSR) linked to a new bacterial blight resistance gene xa33 (t) in rice cultivar 'Ba7'. Maejo Int J Sci Technol 2009, 3:235-247.
26. Korinsak S, Sirithunya P, Meakwatanakarn P, Sarkarung S, Vanavichit A, Toojinda T:Changing allele frequencies associated with specific resistance genes to leaf blast in backcross introgression lines of Khao Dawk Mali 105 developed from a conventional selection program. Field Crops Res 2011, 122(1):32-39.
27. Kullan ARK, Van Dyk MM, Jones N, Kanzler A, Bayley A, Myburg AA, Myburg Z: High-density genetic linkage maps with over 2,400 sequence-anchored DArT markers for genetic dissection in an F_2 pseudo-backcross of Eucalyptus grandis × E. urophylla.
28. Tree Genetics & Genomes 2012, 8:163-175.
29. Lanceras J, Huang ZL, Naivikul O, Vanavichit A, Ruanjaichon V, Tragoonrung S: Mapping of genes for cooking and eating qualities in Thai jasmine rice (KDML105). DNA Res 2000, 7:93-101.
30. Lin XH, Zhang DP, Xie YF, Gao HP, Zhang Q: Identifying and mapping a new gene for bacterial blight resistance in rice based on RFLP markers. Phytopathology 1996, 86:1156-1159.

31. Lu Y, Xiao P, Shao Y, Zhang G, Thanyasiriwat T, Bao J: Development of new markers to genotype the functional SNPs of SSIIa, a gene responsible for gelatinization temperature of rice starch. J Cereal Sci 2010, 52:438-443
32. Luo Y, Yin Z: Marker-assisted breeding of Thai fragrance rice for semi-dwarf phenotype, submergence tolerance and disease resistance to rice blast and bacterial blight. Mol Breeding 2013, 32:709-721.
33. Maclean JL, Dawe DC, Hardy B, Hettel GP: Rice Almanac. 3rd edition. IRRI, WARDA, CIAT and FAO, Philippines; 2002.
34. Manugistics. Statgraphics Plus for Windows 3.0. Manugistics Group, Inc., Rockville: Maryland, USA; 1997.
35. Marchetti MA, Lai X, Bollich CN: Inheritance of resistance to Pyriculariaoryzaein rice cultivar grown in the United States. Phytopathology 1987, 77:799-804.
36. Masouleh AK, Waters DLE, Reinke RF, Ward R, Henry RJ: SNP in starch biosynthesis genes associated with nutritional and functional properties of rice. Sci Rep 2012, 2:557.
37. Molnár S, Galbács Z, Halász G, Hoffmann S, Kiss E, Kozma P, Veres A, Galli Z, Szőke A, Heszky L: Marker assisted selection (MAS) for powdery mildew resistance in a grapevine hybrid family. Vitis 2007, 46(4):212-213.
38. Montoya C, Cochard B, Flori A, Cros D, Lopes R, Cuellar T, Espeout S, Syaputra I, Villeneuve P, Pina M, Ritter E, Leroy T, Billotte N: Genetic architecture of palm oil fatty acid composition in cultivated oil palm (Elaeis guineensis Jacq.) Compared to its wild relative E. oleifera (H.B.K) Cortés. PLoS ONE 2014, 9(5):e95412. doi:10.1371/journal.pone.0095412
39. Noenplab A, Vanavichit A, Toojinda T, Sirithunya P, Tragoonrung S, Sriprakhon S, Vongsaprom C: QTL mapping for leaf and neck blast resistance in KhaoDawk Mali 105 and Jao Horn Nin recombinant inbred lines. Sci Asia 2006, 32(2):133-142
40. Novaes E, Osorio L, Drost DR, Miles BL, Boaventura-Novaes CR, Benedict C, Dervinis C, Yu Q, Sykes R, Davis M, Martin TA, Peter GF, Kirst M: Quantitative genetic analysis of biomass and wood chemistry of Populus under different nitrogen levels. New Phytol 2009, 182:878-890. Ou SH:

41. Kew Surrey: Common wealth Mycological Institute. 2nd edition. 1985.

42. Rajpurohit D, Kumar R, Kumar M, Paul P, Awasthi AA, Basha PO, Puri A, Jhang T, Singh K, Dhaliwal HS: Pyramiding of two bacterial blight resistance and a semi dwarfing gene in type 3 Basmati using marker-assisted selection. Euphytica 2011, 178:111-126.

43. Randhawa HS, Mutti JS, Kidwell K, Morris CF, Chen X, Gill KS: Rapid and targeted introgression of genes into popular wheat cultivars using marker-assisted background selection. PLoS ONE 2009, 4(6):e5752.

44. Roumen E, Levy M, Nottegham JL: Characterization of the European pathogen population of Magnaporthe grisea by DNA fingerprinting and pathotype analysis. European J Pl Path 1997, 103:363-371.

45. Servin B, Martin OC, Mézard M: Toward a theory of marker-assisted gene pyramiding. Genetics 2004, 168:513-523.

46. Siangliw M, Toojinda T, Tragoonrung S, Vanavichit A: Thai jasmine rice carrying QTLch9 is submergence tolerance. Ann Bot 2003, 91:255-261.

47. Siangliw JL, Jongdee B, Pantuwan G, Toojinda T: Developing KDML105 backcross introgression lines using marker-assisted selection for QTLs associated with drought tolerance in rice. ScienceAsia 2007, 33:207-214.

48. Singh A, Singh VK, Singh SP, Pandian RTP, Ellur RK, Singh D, Bhowmick PK, Krishnan SG, Nagarajan M, Vinod KK, Singh UD, Prabhu KV, Sharma TR, Mohapatra T, Singh AK (2012) Molecular breeding for the development of multiple disease resistance in Basmati rice. AoB PLANTS: pls029; doi:10.1093/aobpla/pls029, available online at www.aobplants.oxfordjournals.org.

49. Singh VK, Singh A, Singh SP, Ellur RK, Singh D, Krishnan SG, Bhowmick PK, Nagarajan M, Vinod KK, Singh UD, Mohapatra T, Prabhu KV, Singh AK: Marker-assisted simultaneous but stepwise backcross breeding for pyramiding blast resistance genes Piz5 and Pi54 into an elite Basmati rice restorer line 'PRR78'. Plant Breed 2013, 132:486-495.

50. Sirithunya P, Tragoonrung S, Vanavichit A, Pa-in N, Vongsaprom C, Toojinda T:Quantitative trait loci associated with leaf and neck

blast resistance in recombinant inbred line population of rice (Oryza sativa). DNA Res 2002, 9:79-88.

51. Song WY, Wang GL, Chen LL, Kim HS, Pi LY, Holsten T, Gardner J, Wang B, Zhai WX, Zhu LH, Fauquet C, Ronald PC: A receptor kinase-like protein encoded by the rice disease resistance gene, Xa21. Science 1995, 270:1804-1806.
52. Sreewongchai T, Toojinda T, Thanintorn N, Kosawang C, Vanavichit A, Tharreau D, Sirithunya P: Development of elite indica rice lines with wide spectrum of resistance to Thai blast isolates by pyramiding multiple resistance QTLs. Plant Breed 2010, 129(2):176-180.
53. Suh JP, Roh JH, Cho YC, Han SS, Kim YG, Jena KK: The Pi40 gene for durable resistance to rice blast and molecular analysis of Pi40-advanced backcross breeding lines. Phytopathology 2009, 99:243-250.
54. Suh JP, Jeung JU, Noh TH, Cho YC, Park SH, Park HS, Shin MS, Kim CK, Jena KK:Development of breeding lines with three pyramided resistance genes that confer broad-spectrum bacterial blight resistance and their molecular analysis in rice. Rice 2013, 6:5.
55. Sun L, Su C, Wang C, Zai H, Wan J: Mapping of a major resistance gene to brown planthopper in the rice cultivar Rathu Heenati. Breed Sci 2005, 55:391-396.
56. Sun L, Liu Y, Jiang L, Su C, Wan J: Identification of quantitative trait loci associated with resistance to brown planthopper in the indica rice cultivar Col.5 Thailand. Hereditas 2006, 144:48-52.
57. Sundaram RM, Vishnupriya MR, Biradar SK, Laha GS, Reddy GA, Rani NS, Sarma NP, Sonti RV: Marker assisted introgression of bacterial blight resistance in Samba Mahsuri, an elite indica rice variety. Euphytica 2008, 160:411-422.
58. Tian Z, Qian Q, Liu Q, Yan M, Liu X, Yan C, Liu G, Gao Z, Tang S, Zeng D, Wang Y, Yu J, Gu M, Li J: Allelic diversities in rice starch biosynthesis lead to a diverse array of rice eating and cooking qualities. PNAS 2009, 106(51):21760-21765.
59. Tomita M: Introgression of green revolution sd1 gene into isogenic genome of rice super cultivar Koshihikari to create novel semidwarf cultivar 'Hikarishinseiki' (Koshihikari-sd1). Field Crops Res 2009, 114:173-181.

60. Umemoto T, Aoki N: Single-nucleotide polymorphisms in rice starch synthase IIa that alter starch gelatinization and starch association of the enzyme. Funct Plant Bio 2005, 32:763-768.
61. Umemoto T, Yano M, Satoh H, Shomura A, Nakamura Y: Mapping of a gene responsible for the difference in amylopectin structure between japonica-type and indica-type rice varieties. Theor Appl Genet 2002, 104:1-8
62. Umemoto T, Aoki N, Lin HX, Nakamura Y, Inouchi N, Sato Y, Yano M, Hirabayashi H, Maruyama S: Natural variation in rice starch synthase IIa affects enzyme and starch properties. Funct Plant Bio 2004, 31:671-684.
63. Wang P, Xing YZ, Li ZK, Yu SB: Improving rice yield and quality by QTL pyramiding. Mol Breeding 2012, 29(4):903-913.
64. Watanabe T, Kitagawa H: Photosynthesis and translocation of assimilates in rice plants following phloem feeding by the planthopper Nilaparvata lugens(Homoptera: Delphacidae). J Econ Entomol 2000, 93:1192-1198.
65. Waters DL, Henry RJ, Reinke RF, Fitzgerald MA: Gelatinization temperature of rice explained by polymorphisms in starch synthase. Plant Biotechnol J 2006, 4(1):115-122.
66. Win KM, Korinsak S, Jantaboon J, Siangliw M, Lanceras-Siangliw J, Sirithunya P, Vanavichit A, Toojinda T: Breeding the Thai jasmine rice variety KDML105 for non-age-related broad-spectrum resistance to bacterial blight disease based on combined marker-assisted and phenotypic selection. Field Crops Res 2012, 137:186-194
67. Wongsaprom C, Sirithunya P, Vanavichit A, Pantuwan G, Jongdee B, Sidhiwong N, Lanceras-Siangliw J, Toojinda T: Two introgressed quantitative trait loci confer a broad-spectrum resistance to blast disease in the genetic background of the cultivar RD6 a Thai glutinous jasmine rice. Field Crops Res 2010, 119(2–3):245-251.
68. Xi ZY, He FH, Zeng RZ, Zhang ZM, Ding XH, Li WT, Zhang GQ: Development of a wide population of chromosome single-segment substitution lines in the genetic background of an elite cultivar of rice (Oryza sativa L.). Genome 2006, 49:476-484.
69. Yang Z, Sun X, Wang S, Zhang Q: Genetic and physical mapping of a new gene for bacterial blight resistance in rice. Theor Appl Genet 2003, 106:1467-1472

70. Yi M, Nwe KT, Vanavichit A, Chai-arree W, Toojinda T: Marker assisted backcross breeding to improve cooking quality traits in Myanmar rice cultivar Manawthukha. Field Crops Res 2009, 113:178-186.

71. Yuan HY, Chen XP, Zhu LL, He GC: Identification of genes responsive to brown planthopper Nilaparvata lugens Stål (Homoptera: Delphacidae) feeding in rice. Planta 2005, 221:105-112

72. Zong G, Wang AH, Wang L, Liang GH, Gu MH, Sang T, Han B: A pyramid breeding of eight grain-yield related quantitative trait loci based on marker-assistant and phenotype selection in rice (Oryza sativa L.). J Genet Genom 2012, 39:335-350.

Chapter 6

Planted Forest Development in Australia and New Zealand: Comparative Trends and Future Opportunities

David Rhodes[1] and Michael Stephens[2]

[1]New Zealand Forest Owners Association, Wellington 6143, New Zealand

[2]Australian Forest Products Association,Deakin West ACT 2600, Australia

ABSTRACT

Background

The areas of commercial wood plantations in Australia and New Zealand are 2.01 and 1.71 million hectares respectively, representing the predominant type of planted forest.

Methods

This paper reviews the status of planted forests in Australia and New Zealand and discusses opportunities for further development taking into account economic, social and environmental factors. These factors include regional wood demand and supply conditions, emerging markets for ecosystem services, climate change impacts and social attitudes toward planted forests.

Results

Both countries have followed a similar pattern of state intervention for the development of wood plantations, relying on government ownership and direct incentives such as taxation concessions that have gradually shifted to the use of more indirect policies. The main drivers for wood plantations were as an additional source of wood outside the natural forest, improved productivity and regional development. More recent policy has focused on the privatisation of forest assets and the promotion of ecosystem services such as carbon sequestration. To date, plantings for ecosystem services such as carbon sequestration has been limited.

Conclusions

It is argued that better design of policy settings is needed to capture the carbon abatement benefits and other ecosystem services from planted forests, which may help overcome current investment hurdles. Careful attention will also need to be applied to managing future climate change risks as well as to realising the opportunities from forecast growth in wood and fibre product demand particularly in the Asian region.

INTRODUCTION

The Food and Agriculture Organisation (FAO) (2010) defines a planted forest as those forests 'composed of trees established through planting and/or through deliberate seeding of native or introduced species', which includes forest plantations[a] and planted semi-natural[b] forests. Globally, planted forests have increased in area by just over 8% per cent between 2005 and 2010 to a total of 264 Mha (FAO 2010).

[a]Forest plantations, a subset of all planted forests, are defined as forests of introduced species and in some cases native species, established through planting or seeding, with few species, even spacing and/or even-aged stands (FAO 2006a).

[b]Semi-natural forests are defined as forests of native species, established through planting, seeding or assisted natural regeneration. This definition includes areas under intensive management where native species are used and deliberate efforts are made to increase/optimise the proportion of desirable species, thus leading to changes in the structure and composition of the forest, with possible presence of naturally regenerated trees from other species than those planted/seeded. This may include areas with naturally regenerated trees of introduced species and areas under intensive management where deliberate efforts, such as thinning or fertilising, are made to improve or optimise desirable functions of the forest (FAO 2006b).

Within Oceania, there is approximately 4 Mha of planted forest, with Australia and New Zealand comprising over 92% of this area (Carle and Holgrem 2008). Other countries in the region that have developed a planted forest resource include Fiji and Papua New Guinea. In this paper, planted forest development is discussed with respect to Australia and New Zealand, as the major plantation forested countries in Oceania.

This paper reviews the status of planted forests in Australia and New Zealand and provides a comparative analysis of recent trends and opportunities for their development, taking into account economic, social and environmental factors. These factors include regional wood demand and supply conditions, emerging markets for ecosystem services such as carbon sequestration, climate change impacts and social attitudes toward planted forests. These factors represent a number of challenges and opportunities for planted forest development which to date has focused primarily on wood production. It is argued that better design of policy settings is needed to capture the multiple carbon abatement benefits and other ecosystem services from planted forests, which may help overcome current investment hurdles for new wood plantations. The proximity of Australia and New Zealand to Asia is also discussed in terms of forecast growth in regional demand for forest products and implications for plantation development and associated wood processing.

WOOD PLANTATIONS

Planted forests currently play an important role in Australia and New Zealand as a source of wood and fibre supply, partly in response to the declining availability of industrial wood from natural forests. The plantation estates of New Zealand and Australia produce roughly the same volume of logs per annum at 26 million and 27 million cubic metres respectively. In terms of their relative contribution to total annual log supply, wood plantations contribute 99.9% in New Zealand and over 80% in Australia (Australian Bureau of Agricultural and Resource Economics and Sciences (ABARES) 2014; New Zealand Forest Owners Association 2013). This reflects a contraction over time in natural forest availability for wood production and a maturing plantation resource for harvest.

In Australia, the area of commercial wood plantation is just over 2 Mha, with 1.02 Mha of coniferous plantation and 980 000 ha of broadleaved plantation (Gavran et al. 2012). Australia's commercial plantations are located mainly in the north-eastern coastal, south-western, southern and eastern regions of Australia.

The coniferous plantation is dominated by *Pinus radiata* D. Don or radiata pine (75%) and *Pinus elliottii* Engelm. and *Pinus caribaea* Morlet hybrids known as southern pines (15%). Most radiata pine is located in Victoria, New South Wales and Tasmania, and most southern pines are located in south east Queensland. *Pinus pinaster* Aiton, or maritime pine, is the dominant coniferous plantation species in Western Australia. The broadleaved plantation is dominated by *Eucalyptus globulus* Labill. (55%) and *Eucalyptus nitens* H. Deane & Maiden (24%), which are primarily managed for short-rotation pulpwood production.

In the seasonally dry tropical regions of northern Australia, there is also around 10 000 ha of planted African mahogany (*Khaya senegalensis* (Desr.) A.Juss.) which has considerable potential as a high value timber plantation species (Dickinson et al. 2011), as well as over 7 600 ha of Indian sandalwood (*Santalum album* L.) which is used in the global fragrance market.

In New Zealand, there is 1.71 Mha of wood plantation which is dominated by exotic species representing 99% of total planted area, including radiata pine (90%) and *Pseudotsuga menziesii* or Douglas-fir (6%). The remainder of the plantation estate is comprised of a mixture of Cypress species and other exotic softwoods and hardwoods (New Zealand Forest Owners Association 2013).

HISTORICAL DEVELOPMENT

Planted forest development in Australia and New Zealand has followed a similar pattern in terms of direct state intervention for the development of wood plantations over the last 150 years or so, which has gradually shifted to more indirect policies for the improved operation of markets and the promotion of ecosystem services such as carbon sequestration. These policies have resulted in a sizeable wood plantation resource and processing industry, whereby the forest products industry contributes $NZ4.5 billion and $A6.3 billion to national gross domestic product (GDP) in New Zealand and Australia respectively. The primary focus on wood production reflects international trends for planted forests, where it is estimated that 76% of global planted forests have production as their primary goal (FAO 2010).

Stephens and Grist (2014) and Ferguson (2014a) describe early state intervention in Australia for wood plantations via direct government ownership that has shifted to indirect policies such as the Plantations 2020 Vision. The Plantations for Australia: 2020 Vision, launched by the federal and state governments in 1997, aimed to enhance wealth creation and international competitiveness through a notional target of trebling the area of wood plantations to 3 Mha by 2020. This policy has focused on improving competitiveness through research and development, improving information on markets and prices and streamlining planning and regulation.

Large-scale wood plantation development began in the 1960s, through the use of concessional loans from the federal government to the state governments which established and managed the development of long-rotation (e.g. 35 year) coniferous plantation. The major goals of this expansion were to promote regional development and increase wood productivity and self-sufficiency, given the adverse terms of trade in forest products at that time (Ferguson 2014a). These loans led to the establishment of 1 Mha of coniferous plantation for integrated sawlog and pulpwood, used for domestic sawn timber and pulp and paper production (Figure 1). The other major trend that occurred from the 1990s was the emergence of Managed Investment Schemes (MIS) which took advantage of existing taxation incentive provisions and pooled capital from a large number of small investors. This led to the rapid rise of mostly short rotation hardwood (*Eucalyptus* spp.) plantations for woodchip export markets, which increased from a relatively low base to over 800 000 ha by 2012.

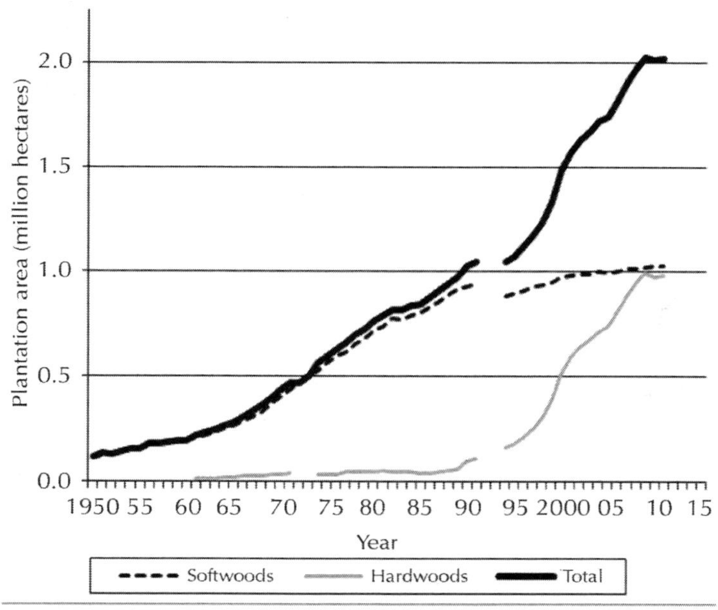

Figure 1: Plantation forest area in Australia, 1950-2011 (*Source*: ABARES.)

However, problems with the structure of the financial arrangements and issues with access to suitable land led to the collapse of many MIS companies, with some areas planted on uneconomic sites. Following the onset of the Global Financial Crisis in 2007-8 and the collapse of many MIS companies, most of this estate has been acquired by institutional investors. Further rationalisation of the former MIS estate is projected, with some areas not expected to be commercially replanted as a consequence of being established on poor quality sites or other factors such as long distances to markets. Some companies estimate that up to 30% of the former MIS hardwood plantation estate may not be replanted (Ferguson 2014b), with this land being converted back to agriculture following harvest (Stephens and Grist 2014).

Other forests have been planted for ecosystem services such as soil and water conservation, biodiversity and carbon sequestration. There is estimated to be 153 000 ha of non-industrial planted forest in Australia, some of which is used for environmental purposes and farm forestry activities (Montreal Process Implementation Group for Australia and National Forest Inventory Steering Committee 2013). Consultants (URS Forestry 2008) reported 155 000 ha of farm forestry plantings in Australia based on a 2008 survey, which included some large-scale plantings as part of leasehold arrangements with MIS companies. In addition to wood and agricultural productivity benefits (e.g. shade, shelter), a proportion of these plantings is for ecosystem services such as biodiversity and soil conservation, consistent with earlier surveys of farmers undertaking tree planting activities (Wilson et al. 1995).

Similarly, in New Zealand, there was a period of direct state intervention for wood plantation promotion from the 1960s through to the early 1980s. Over this period the plantation forest estate expanded by almost 700 000 ha through a range of programmes, including forestry encouragement grants, low interest loans and taxation deductions of up to 100% of planting costs to support private investment, as well as government planting programmes through the New Zealand Forest Service (Rhodes and Novis 2002, de Fegeley et al. 2011).

Other priorities for planted forests in New Zealand have included land rehabilitation and control of soil and water erosion from previous vegetation clearing for agriculture, particularly on steep slopes. In the Gisborne district of the North Island, for example, there has been a history of planting to combat erosion through Government-owned plantations and grants to private landowners. This has included the East Coast Forestry Project, which has assisted landowners to plant trees since 1992 for multiple goals including erosion control, wood production and employment (Rhodes 2001). Reforestation has been used to improve water quality that declined due to eutrophication from intensive agriculture (Maclaren 1996), and more recently, for the development of a carbon trading scheme.

Ownership Trends

A significant feature of the wood plantation estate in Australia has been the shift in ownership from the public sector (i.e. state governments) to the private sector (Figure 2). The harvest and management rights to most of the coniferous plantations were sold in the 2000s to private institutional investors as part of the commercialisation of government owned assets. The management of these plantations is undertaken by timber investment management organisations (TIMOs) on behalf of investors, consistent with recent trends in North America. The other trend has been the acquisition by TIMOs of the hardwood eucalypt plantations formerly owned or managed by the failed MIS companies.

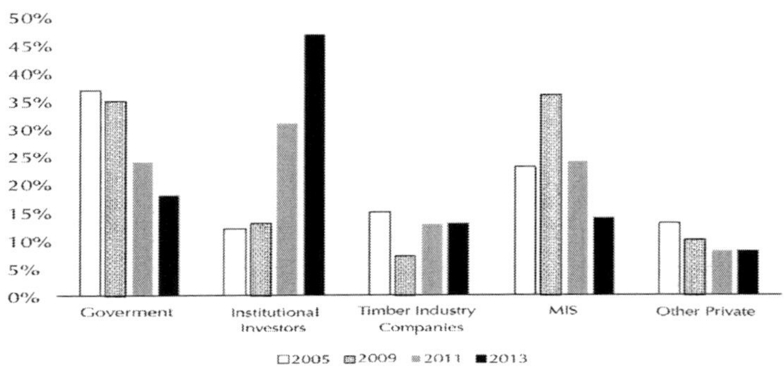

Figure 2: Plantation ownership in Australia (*Source*: ABARES; Australian Forest Products Association).

In New Zealand, there is 1.71 Mha of wood plantation. Fifteen companies manage plantation forest in excess of 20 000 ha with the largest being just over 235 000 ha (New Zealand Forest Owners Association 2013). These fifteen companies manage 62% of the plantation estate. Six companies identify themselves as TIMOs, managing 389 000 ha, and 13 companies identify as property management companies managing just over 1 Mha. Maori and the Crown own considerably more land than forest, with Maori leasing 326 000 ha and the Crown leasing 219 000 ha to forest management companies. Another feature of New Zealand's ownership pattern is that 629 000 ha (or just over 37%) is owned by private individuals and landowners with less than 10 000 ha. An increasing proportion of this area will reach harvest age in the next 5 to 10 years.

REGIONAL WOOD DEMAND AND SUPPLY

Given their close proximity and on-going forest products trade in Asia, wood demand and supply in the region has a direct influence on profitability and investment in wood plantations in Australia and New Zealand. There is also considerable intra-regional trade between the two countries. Australia is a net importer in forest products of around $2 billion per annum, comprising mostly imports of paper and paperboard from China and sawn timber from New Zealand. Australia has also been a large exporter of hardwood woodchips to Japan and China for pulp and paper production, which has averaged over $A850 million per annum over the past decade. Log exports to China have also increased markedly from 201 000 m^3 in 2002-03 to over 1.2 million m^3 in 2012-13 (ABARES 2014).

New Zealand is a large exporter of forest products, worth around $NZ 4.5 billion in 2012, reflecting the importance of the sector to the national economy at 3.3% of GDP (New Zealand Forest Owners Association 2013). In 2012, the major forest product export partners by value included China ($1.45 billion), Australia ($775 m), Japan ($542 m) and Korea ($392 m).

In terms of demand for forest products, China's influence is increasingly significant for Australia and New Zealand. Over the past few decades, woodchip exports to Japan have been the major market for Australian hardwood plantation pulpwood. However, China's share of woodchip exports value from Australia has grown from less than 1% in 2001-02 to around 20% in 2011-12 (ABARES 2014). There has also been a fundamental shift in overseas market share for New Zealand's forest products, particularly the rise of China to 67% of log and pole exports in 2012. In 2012, China represented 34% of New Zealand's total forest product exports by value, compared to just 4% in 1992 when Australia's share was 34% of forest product exports.

The demand outlook in Asia is influenced by economic growth assumptions and underlying social and environmental policies. Buongiorno et al. (2012) and Indufor (2012) present a range of global demand scenarios in terms of population growth, GDP and adoption of environmental and climate change policies. A consistent trend in these scenarios is the high economic growth in China and India over the next few decades and their forecast position as net forest product importers. TheAsia Pacific Forestry Commission (2010) forecast demand for industrial round wood to increase from 317 million m^3 in 2005 to 550 million m^3 in 2020, with East Asia, especially China, accounting for most of this consumption. Similar analysis has shown that growth in construction is expected to be strong in China, the United States and India, with a 70% increase in construction output to $15 trillion worldwide by 2025, including the need for 270 million new homes in China and India by 2015 (Global Construction Perspectives and Oxford Economics 2013).

Global wood pellet production, as a source of fuel for energy, has grown from just less than 2 Mt in 2000 to around 16 Mt in 2010 (Lamers et al. 2012). Global wood pellet demand is expected to grow strongly to around 40-50 Mt per annum by 2020 with Europe accounting for a large share of consumption. The Asian region, which is a much closer market than Europe for Australia and New Zealand, is expected to rise from less than 1 Mt in 2010 to around 10-15 Mt by 2020 (International Energy Agency 2011).

The outlook for Australian and New Zealand wood plantations is also affected by local demand, which is largely influenced by Australian economic activity as a medium sized economy of almost 23 million people (Australian Bureau of Statistics 2014). Recent forecasts of domestic consumption suggest a moderate increase for most forest product categories (Table 1), based on assumed economic growth, rates of household formation, population growth and substitution of forest products in final demand.

Table 1: Long-term consumption forecasts: Australia

Time period				
Product	**2010-11** a	**2012-13** a	**2029-30**	**2049-50**
Sawn wood (Mm^3)	5.0	5.1	6.0	6.5
Wood based panels (Mm^3)	2.0	1.7	3.1	4.3
Paper and paperboard (Mt)	4.0	3.6	5.8	7.1

Source: ABARES (2013, 2014). a = actual consumption. Mm3 = million cubic metres. Mt = million tonnes.

Rhodes and Stephens New Zealand Journal of Forestry Science 2014 44(Suppl 1):S10 doi:10.1186/1179-5395-44-S1-S10

The consumption of forest products in New Zealand has similarly remained steady over the past few years, albeit from a smaller population base than Australia (Table 2). As a net exporter of forest products with a relatively small domestic market, growth opportunities primarily lie in the global market.

Table 2: Consumption of forest products: New Zealand

Year			
Product	**2006**	**2010**	**2014**
Sawn wood (Mm^3)	2.5	1.9	2.2
Wood based panels (Mm^3)	0.7	0.4	0.8
Paper and paperboard (Mt)	0.7	0.9	0.8

Source: Ministry for Primary Industries (2014).

Rhodes and Stephens New Zealand Journal of Forestry Science 2014 44(Suppl 1):S10 doi:10.1186/1179-5395-44-S1-S10

On the supply side, Carle and Holgrem (2008) undertook a global survey of planted forests in 61 countries and predicted around a one third increase in total area over the next two decades, to reach 345 Mha by 2030 (Table 3). By comparison, Oceania was predicted to have an increase in area of around 11% or 400 000 ha by 2030.

Table 3: World and Oceania planted forest area, 2005 (actual) and 2030 (forecast)

Year	Region	Softwoods				Hardwoods						Total
		***Pinus* spp.**		**Other spp.**		***Acacia* spp.**		***Eucalyptus* spp.**		**Other spp.**		
2005	Oceania	**Mha**	%	**Mha**	%	Mha	%	Mha	%	Mha	%	Mha
		2.7	68	0.2	5	0.0	0	0.5	13	0.2	5	3.6
	World	73	28	64	25	9	0.3	14	5	100	38	261
2030	Oceania	2.9	69	0.2	5	0.0	0	0.8	19	0.3	7	4.2
	World	94.9	28	83.0	24	9.9	0.3	21.4	6	135.5	39	344.6

Source: Carle & Holgrem (2008). Note: Mha = millions of hectares.

Rhodes and Stephens New Zealand Journal of Forestry Science 2014 44(Suppl 1):S10 doi:10.1186/1179-5395-44-S1-S10

According to Indufor (2012), the total area of industrial forest plantation in the world was 54.3 Mha in 2012 and may increase to 91 Mha by 2050. Indufor (2012) estimated that the area of industrial plantation in Asia and Latin America could double by 2050 with some expansion in Africa and Oceania.

In Australia, a recent survey of plantations and log availability has estimated that the plantation area may remain relatively constant at 2 Mha up to 2019-20 (Gavran et al. 2012). However, log volumes are expected to increase as a result of the maturing resource to average over 29 million m^3 per year in 2015-19 and reach an average annual peak of 33 million m^3 in 2030-34, largely as a result of an increase in hardwood pulpwood with softwood volumes to remain relatively static (Figure 3).

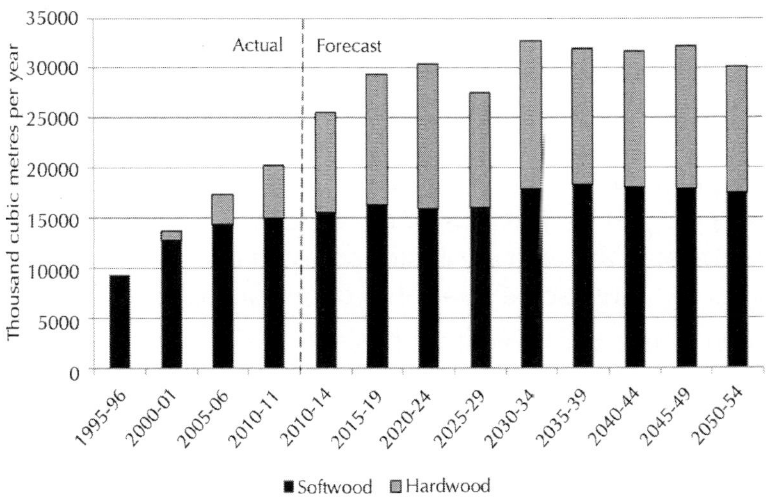

Figure 3: Australian plantation log availability 1990-2050 (*Source:*Gavran et al. (2012); ABARES (2014)).

While providing a useful basis for national wood assessments, Ferguson (2014b) has identified some limitations in plantation forecasting in Australia, including the need for more robust sampling for growth and yield and information on replanting assumptions. This is particularly relevant in the case of the rationalisation of the former MIS plantations, with some areas not to be replanted for the reasons identified above.

In New Zealand, the harvest level is around 30 million m^3 of wood from plantations (Ministry for Primary Industries, 2014). Although market conditions eased in 2014, the steady increase in harvest volumes over the past five years is forecast to continue in the coming years to reach around 35 million m^3 by 2022 (Wood Council of New Zealand 2012). Most of this increase is to come from small growers who established forests in the early 1990s and actual harvest levels will depend upon market conditions and the decisions of small-scale owners (Ministry for Primary Industries 2013a).

ECOSYSTEM SERVICES

The predominant purpose of global planted forests has been for wood and fibre production (FAO 2010), which has also been the case in Australia and New Zealand. However, it is expected that the relative importance of planted forests for environmental purposes will increase in the future, given their potential to contribute to a range of ecosystem services such as rehabilitation of degraded sites, protection of soil and water, regulation of watersheds, biodiversity and carbon sequestration (Bahaus et al. 2012). This has been the case in China, for example, which has tripled the area of planted forest between 1990 and 2010 to 60 Mha for the large scale control of desertification (FAO 2010).

Planted forests can have both positive and negative environmental consequences such as the introduction of pests and water-use impacts, but through careful design and planning can help address landscape management issues (Nambiar and Ferguson 2005). When assessing the scope for forest plantings in Australia and New Zealand for ecosystem services, it is important to note the previous environmental degradation of land, water and vegetation, primarily as a result of agricultural development and other settlement patterns since the arrival of Europeans in the late 18th century. In Australia, it is estimated that around 13% or 100 Mha of land has been cleared of native vegetation, mostly eucalypt woodlands, with a further 65% subject to disturbance pressures such as livestock grazing (State of the Environment Committee 2011). This has led to a range of natural-resource management problems, including soil and water erosion, sedimentation, acidification, dry-land salinity and loss of plant and animal diversity.

Plantings for ecosystem services in Australia have evolved from government sponsored schemes (e.g. grants, community education) in the 1980s for land rehabilitation and biodiversity to broader climate change mitigation schemes. Stephens and Grist (2014) identified three distinct trends for ecosystem services plantings, comprising:

- programmes for landscape restoration and farm productivity (1980s);
- greater recognition of biodiversity values and landscape connectivity from revegetation and planted forests (1990s); and
- the emergence of carbon markets for forestry and land sector activities (2000s).

In both Australia and New Zealand, research has found that plantations can provide habitat benefits for a number of species compared to agricultural and pastoral land (Maclaren 1996, Lindenmayer and Hobbs 2004, Kavanaugh et al. 2007, Hock et al. 2009, Baral et al. 2013). There has also been considerable interest in the development of market-based mechanisms to capture ecosystem services such as biodiversity for forest owners. Keenan et al. (2004) identified the early potential for market based mechanisms to encourage small scale private landowners in Australia to deliver environmental services, such as to retain existing forest, establish forests on previously cleared land and integrate trees more fully with farming systems. It was argued that while forest owners can make a profit from the sale of wood and other commercial products, environmental services were not effectively valued or traded as market goods. However, it has been difficult for plantation owners and investors to generally capture the non-market benefits from ecosystem services, given the lack of market development and range of site-specific outcomes and conditions.

In a dry continent such as Australia, it is recognised that plantations, like all forms of agricultural land use, intercept water and that plantations generally use more water than pasture (Zhang et al. 2003). Plantations can also play an important role in improving water quality through reduced use of chemicals and pesticides, reduced erosion and soil salinity and flood control (Zhang et al. 2007). Water policy in Australia has therefore been directed at achieving an appropriate balance between managing the impacts of plantations on both water yield and quality. The interaction of plantations on both these variables is important, as the costs and benefits of reduced run-off and/or lower water tables will depend on the salinity of the groundwater, its potential to contribute to waterlogging or inundation and its potential for productive use (Zhang et al. 2007). In the south-east region of South Australia, for example, groundwater is a highly valued freshwater resource, while in Western Australia there is high saline groundwater and dry-land salinity issues from previous land clearing. At a national level, dry-land salinity is a particularly important water management issue, which arises from the removal of vegetation and rising saline water tables that causes land to be unproductive. The National Land and Water Resources Audit (2001) estimates that 5.7 Mha of land is adversely affected by dry-land salinity which is forecast to increase to 17 Mha by 2050 in the absence of amelioration.

Zhang et al. (2007) suggest that afforestation could play a significant role to reduce dry-land salinity, given the ability of trees to manage recharge through their deep rooting habit and high perennial water use. Considerable work has been undertaken in assessing and trialling salt tolerant tree species to ameliorate dry-land salinity in a range of geographic regions and sites, which can also produce other related benefits such as the harvest of biomass for fodder or bioenergy (Dale and Dieters 2007, Sochaki 2012) and deliver multiple wood, water quality and carbon sequestration benefits (Townsend et al. 2012).

Stephens and Grist (2014) reviewed the market potential for delivering joint wood production and ecosystem services from planted forests in Australia, including biodiversity, water quality and carbon sequestration. They concluded that there was limited scope, at least in the short term, for biodiversity payments given the complexity of capturing the benefits and transaction costs. By comparison, dry-land salinity mitigation was considered important in a regional context where markets do exist at a catchment scale and there is greater market development as part of the national water market for consumptive uses. Carbon markets for sequestration were considered most prospective, given the development of a domestic crediting and verification mechanism and the fungibility of tradable carbon credits.

In New Zealand, Hock et al. (2009) assessed the potential for ecosystem services from planted forests and reached similar conclusions regarding a lack of information and tradable markets for services other than for carbon sequestration. Despite commercial limitations to capturing ecosystem services, Yao et al. (2014) conducted a choice-modelling experiment and found that biodiversity enhancement in large-scale planted forests is valued in the community, and could potentially form part of a suite of future goods and services.

Carbon Markets

Planted forests can provide multiple carbon sequestration and fossil fuel displacement benefits, through the carbon stored in forests and harvested wood products, the substitution of more energy-intensive products with wood products and the use of woody biomass for renewable energy (Malsheimer et al. 2011, Lippke et al. 2011, Oliver et al. 2014).

Both New Zealand and Australia have been early movers in terms of the implementation of domestic carbon market based schemes with mixed results in terms of promoting new plantings. In Australia, this has included the New South Wales Greenhouse Gas Reduction Scheme (GGAS) and the Carbon Farming Initiative, the latter which links as a land crediting mechanism to the national Emissions Reduction Fund (ERF). The ERF is a reverse auction market mechanism to buy low-cost emission reduction activities in the Australian economy, with the Australian Government initially allocating $1.55 billion to the Fund over a three year period commencing 1 July 2014 (Commonwealth of Australia 2013).

Mitchell et al. (2012) provided a summary of the evolution of carbon forestry schemes and their uptake in Australia, which has resulted in the establishment of 65 000 ha. The main schemes have generally been aimed at Kyoto Protocol compliant activities, such as afforestation and reforestation activities on cleared agricultural land post-1990. Most carbon plantings have been for not-for-harvest projects, either mixed plantings of eucalypt species, or plantings of mallee eucalypts in low rainfall areas that are multi-branched trees with a height rarely exceeding 6 metres. Overall, the uptake of wood plantations has been minimal given uncertainty in carbon market settings and scheme design issues (Mitchell et al. 2012). These issues have included the treatment of additionality, carbon pools (e.g. inclusion of wood products) and crediting periods (Stephens and Grist 2014). For example, the proposed design of the ERF is to have five-year contracts, which may disadvantage land based sequestration such as forestry projects, which typically have low early growth but higher longer term sequestration rates.

However, given the large area of previously cleared land, there has been strong interest in the potential for sequestration from planted forests under a range of carbon price and land availability scenarios. Several spatial analyses at a national scale have assessed the economic potential for plantations under a carbon market using physiological growth models and agricultural land opportunity costs, with estimates ranging from less than 350 000 ha up to 9 Mha by 2050 (Stephens and Grist 2014). In the absence of a carbon price, Paul et al. (2013) found that establishment of new plantations for wood production on agricultural land was not presently viable in Australia given current costs of establishment. However, on average, for scenarios of hardwoods (*Eucalyptus* spp.) and softwoods (e.g. *Pinus radiata*), a payment of about \$A10 and \$A30/t CO_2 respectively would make plantation expansion viable assuming a 4% carbon price increase per year, with an initial price above \$A20/t CO_2 needed in the majority of scenarios for south-eastern Australia. The scenarios included short- and long-rotation plantations for pulpwood and sawlogs and the crediting of carbon stored in wood products and soils. The findings are consistent with de Fegeley et al. (2011) who suggested that without a carbon price new wood plantations would not be able to meet the investment hurdle rate.

In New Zealand, an emissions trading scheme was introduced in 2009 and forestry was the first sector to be included, generating an initial stimulus of forest plantings for sequestration. In 2011, for example, new forest plantings were almost double the previous year at 12 000 ha (New Zealand Forest Owners Association 2013). However, scheme design issues have since eroded the incentives for planted forests through low domestic carbon prices and high volatility and risk brought about by a series of amendments to the scheme. The depressed New Zealand carbon prices are attributed to the partial sectoral coverage of the scheme and strong linkage to uncapped international carbon trading of Kyoto compliant units, which have also fallen significantly in recent years. Under the New Zealand scheme, any credits are also balanced against any liability from the net release of carbon at harvest.

The ability of a carbon price to drive land use change is also dependent on the agricultural revenues foregone and the objectives of the owner (West et al. 2011). Manley and Maclaren (2009) estimate that a price of \$NZ20/t CO_2 could improve the internal rate of return on a typical radiata pine plantation from 5.4% to 11.0% and could stimulate up to 25 000 ha/annum of new planting. In addition, carbon prices can result in modified management practices in new and existing planted forests. Hale et al. (2014) modelled the trade-offs between timber harvest and carbon sequestration and found that while more productive regions of New Zealand such as the Central North Island and Northland have a greater capacity as a carbon sink, it is the less productive regions that have a comparative advantage in carbon sequestration in terms of a lower cost of wood production revenue forgone. Various treatments such as switching from pruning to plant-and-leave or other harvest constraints may be optimal for a range of assumed carbon prices.

Forest Management

The legal and institutional policy framework for sustainable forest management in Australia and New Zealand is well described in McDermott et al. (2010) and includes government environmental and planning regulation and codes of forest practice for plantation operations. These codes and regulations prescribe operational practices such as stream buffer areas that cannot be planted, how to build and maintain roads, use of fertilisers and pesticides, erosion control and fire protection requirements. In addition, there is high adoption of voluntary internationally recognised certification schemes for sustainable forest management. In Australia, around three-quarters of production forests are certified under either, or both, the Forest Stewardship Council (FSC) or the Programme for the Endorsement of Forest Certification (PEFC) via the Australian Forestry Standard. Similarly, New Zealand plantation forests began obtaining FSC certification during the mid-1990s and almost 60% of New Zealand's planted forests are certified under FSC.

CLIMATE CHANGE IMPACTS

Climate change can have an important impact on planted forest development. Several studies have looked at the possible impacts of climate change on planted forests in Australia and New Zealand (Watt et al. 2008, Battaglia et al. 2009, Booth et al. 2010, Singh et al. 2010, ABARES 2011), including the effects of predicted changes in CO_2

concentration, temperature and rainfall, as well as the threats posed by pests, diseases, weeds, fire and drought.

Climate forecasts vary, but generally predict that New Zealand could be 2°C warmer from 1990 to 2090 and Australia could be warmer by between 0.4 and 2°C by 2030 and between 1 and 6°C by 2070. Variable rainfall trends are also predicted, with the western side of New Zealand's two large islands predicted to have increasing rainfall while the eastern side less (Watt et al. 2008). In Australia, it is predicted that southern Australia will have less rainfall with some increases in northern Australia.

A key consideration is the impact of climate change on forest growth. Plantation growth may increase from rising levels of atmospheric carbon dioxide but may decrease from rising temperature and increased water loss (evapotranspiration). Increased temperature may also increase productivity in some regions and for some species, such as an extended growing season in higher altitude regions and reduced frost. Battaglia et al. (2009) modelled the effects of climate forecasts on plantation productivity in Australia and found a net increase in growth for some major growing regions and for other regions a net decrease. Drought-induced tree mortality was also identified as a significant risk in some drought-prone regions. Watt et al. (2008) identified a general gain in forest growth in New Zealand given relatively high water availability but increased drought risk in some areas.

The other major consideration is the impact of the secondary effects of climate change. In New Zealand, there is forecast to be stronger winds and more high intensity rainfall as well as less risk of fungal disease in drier areas but increased risk from fire and insect attack. Fire risk is also forecast to be significant in Australia. Hennessy et al. (2007) predicted that the frequency of very high and extreme fire danger days in Australia is likely to rise 4-25% by 2020 and 15-70% by 2050. Stephens (2010) has identified that these risks raise significant fire management issues at an individual stand and landscape scale, considering that planted forests occupy less than 2% of the total forest area in Australia of 125 Mha. Cyclones in northern Australia are also predicted to increase in frequency and impact on areas further south of current ranges. Higher susceptibility to pest and disease outbreaks is predicted, resulting from changes in both pathogen vectors and the health of trees influenced by climatic factors, although these factors are the least understood in both Australia and New Zealand (Watt et al. 2008, Singh et al. 2010).

Overall, Australian plantation forests are classified as moderately vulnerable to climate change given that the major commercial species are grown over wide climatic ranges (Booth et al. 2010). It is also recognised that there is large capacity for adaptation. The main management responses are described by Booth et al. (2010), including: spacing and thinning, watering and nutrient regimes, fire management, pest, disease and weed management, use of genotypes and site selection, climate monitoring and establishment practices. Adaptive forest management is regarded as an important tool for plantation managers and the industry has responded in Australia through the development of capacity building tools such as a plantation forest industry climate change adaptation handbook (Stephens et al. 2012).

SOCIAL ATTITUDES AND LAND ACCESS

Commercial afforestation of agricultural land is often associated with social conflict over the perceived environmental, economic and social impacts of plantations (Schirmer 2007). Planted forest development in Australia and New Zealand has attracted some community concerns, primarily regarding competition for agricultural land, employment impacts and environmental issues. Williams (2008) undertook a survey of community attitudes to wood plantations in Tasmania and the south-western regions of Western Australia and found that respondents tended to value agriculture land use higher than for plantation forestry. Many people believed plantations offered benefits to companies and individuals but only limited regional employment and economic benefits. Similarly, Langer and Barnard (2002) surveyed residents in the Gisborne East Coast region of New Zealand and identified that community concerns about forest expansion included lack of employment generation and the perceived loss of good farming land. These views have been exacerbated in Australia from the collapse of many MIS companies and concerns over the rapid expansion of plantations prior to their collapse and upward pressure on land prices in some areas (Ferguson 2014a). Other issues have included perceptions of naturalness and implications for planted forest management in New Zealand (Fairweather and Swaffield 2003).

Research has shown that plantation forestry can generate a range of employment and economic benefits in Australian regions, particularly through integrated growing, harvesting, transport and processing activities (Bureau of Rural Sciences 2005).

These issues are important as any expansion of the plantation estate will require access to rural land and engagement with landowners and communities. A related issue is the relationship between large-scale and small-scale plantation forestry where the latter are generally viewed more positively (Schirmer 2007). This has led to greater policy recognition of the need for improved relationships between the plantation industry and the rural sector to achieve an appropriate mix of industrial scale and small scale plantings (de Fegely et al. 2011, Ferguson 2014a). To this end, Gordon et al. (2013) assessed the extent of community engagement by the Australian plantation forest industry, and while stakeholders often considered the engagement to be ineffective, measures were identified to enhance relationships with stakeholders and the required skills of forestry professionals.

DISCUSSION

Internationally, the development of wood plantations has struggled to attract large-scale private investment without some form of government intervention or subsidy (Nielson 2007). This is primarily because of the time value of money with high up-front establishment costs and a long period until harvest returns. This, in turn, results in a lower rate of return for plantations compared with other investments excepting where there is other comparative advantage (e.g. high product prices, low transport costs, high yields). This has been the case in Australia, and to some extent New Zealand, where the high opportunity cost of land for agriculture has inhibited new investment without some form of government support. This is reflected in the cumulative plantation forest area in Australia over time (Figure 1), where early direct intervention (i.e. state ownership) and assistance schemes generated significant new plantings from the early 1960s.

However, it is important that development policies are well-targeted to address market failure, minimise distortions and promote the long-term competitiveness of plantations. This is particularly relevant in terms of the potential over-stimulation of the sector through previous subsidy experiences in New Zealand (Rhodes and Novis 2002), and more recent uptake of taxation measures in Australia. In Australia and New Zealand, government support has now transitioned to more indirect enabling policies, with a recent decline in new plantings partly as a consequence of market conditions (Figure 4).

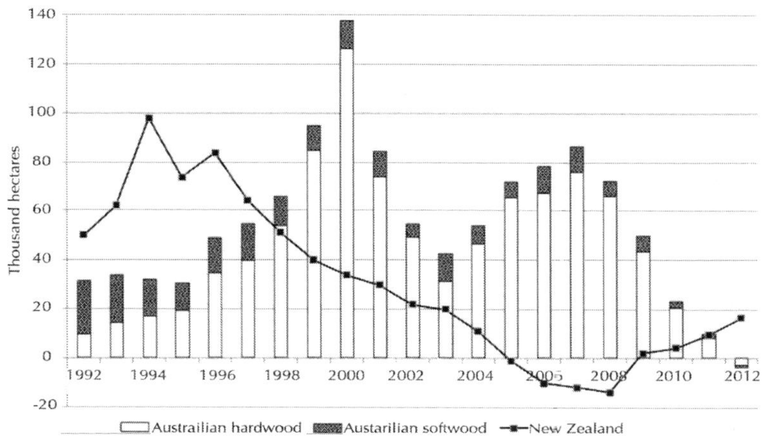

Figure 4: New plantations established in Australia and New Zealand (*Source*: ABARES and New Zealand Forest Owners Association. Note: new plantation areas are net of plantation conversion to other land use).

A key trend is the sharp decline in hardwood (i.e. eucalypt) planting rates since 2007 in Australia and plateauing or fall in softwood plantations in both Australia and New Zealand since the 1990s. This will act as a constraint to the processing industry's ambitions to expand the scale of processing to achieve internationally cost-competitive levels using local wood and fibre (de Fegely et al. 2011).

On the export side, plantation owners in Australia and New Zealand have taken advantage of increasing demand for both softwood and hardwood logs and woodchips in China. With respect to the eucalypt plantation resource, Australian producers have experienced a recent upswing in exports to China, although price competitiveness with other emerging regional suppliers such as Vietnam will be important. In June 2014, Vietnam took over from Australia as the world's largest supplier of woodchips to pulp mills in Asia (Pulp and Paper News 2014).

The outlook for plantation forest product exports to Asia from Australia and New Zealand will be influenced by the competitiveness of other major suppliers. The expansion of fast-growing wood plantations (i.e. rotations less than 40 years) in many parts of the world represents a challenge in terms of improved growth (measured as mean annual increment (MAI)) (Table 4) and increased supply. While growth rates in Australia and New Zealand are comparable to many other countries, there can be considerable variation as in the case of Australia (Table 5). Growth rates vary depending on tree species and region, reflecting underlying differences in climatic and edaphic conditions.

Table 4: Selected international growth rates for wood plantations

Species	**Mean Annual Increment (m^3 ha^{-1} yr^{-1})**	**Rotation length (years)**	**Main countries or regions**
Eucalyptus grandis W.Hill ex Maiden and various hybrids	15 - 40	5 - 15	Brazil, South Africa, Uruguay, India, Congo, Zimbabwe
Other tropical eucalypts	10 - 20	5 - 10	China, India, Thailand, Vietnam, Madagascar, Myanmar
Temperate eucalypts	5 - 18	10 - 15	Chile, Portugal, north-west Spain, Argentina, Uruguay, South Africa, Australia
Tropical acacias	15 - 30	7 - 10	Indonesia, China, Malaysia, Vietnam, India, Philippines, Thailand
Caribbean pines	8 - 20	10 - 18	Venezuela
Pinus patula Schiede ex Schltdl. & Cham. and *P. elliotii*	15 - 25	15 - 18	Swaziland
Gmelina arborea Roxb.	12 - 35	12 - 20	Costa Rica, Malaysia, Solomon Islands
Paraserianthes falcataria (L.) Nielsen	15 - 35	12 - 20	Indonesia, Malaysia, Philippines

Poplars	11 - 30	7 - 15	China, India, USA, central and western Europe, Turkey
Pinus radiata *	25 - 30	25	Australia, New Zealand, South Africa, Chile

Source: Cossalter and Pye-Smith (2003), * FAO (2001).

Rhodes and Stephens New Zealand Journal of Forestry Science 2014 44(Suppl 1):S10 doi:10.1186/1179-5395-44-S1-S10

Table 5: Australian regional growth rates for wood plantations

Species	**Mean Annual Increment** **(m^3 ha^{-1} yr^{1})**	**Rotation length** **(years)**	**Region(s)**
Temperate eucalypts (pulpwood)	10 - 19	12 - 25	Tasmania, South Australia, New South Wales, Victoria, Queensland, south-west Western Australia
Temperate eucalypts (sawlog)	20 - 27	14 - 20	Tasmania, South Australia, New South Wales, Victoria, Queensland, south-west Western Australia
Araucaria spp.	15	50	Queensland, northern New South Wales
Pinus radiata	16 - 21	30	Tasmania, Victoria, New South Wales, south-west Western Australia, South Australia
Pinus elliotii and *P. caribaea*hybrids	15 - 18	30	Queensland, northern New South Wales
Eucalyptus nitens	15	15	Tasmania

Source: Gavran et al. (2012).

Rhodes and Stephens New Zealand Journal of Forestry Science 2014 44(Suppl 1):S10 doi:10.1186/1179-5395-44-S1-S10

However, the long-term sustainability, expansion and productivity of many international wood plantations are uncertain, reflecting underlying environmental and social issues. These issues include population pressure, small land holdings, technology transfer, environmental practices and competition for land and food production (Cossalter and Pye-Smith 2003, Kugelman and Levenstein 2012). The Asia Pacific Forestry Commission (2010) noted that while plantation productivity may increase, this will depend in many countries on improving favourable policy and institutional environments. In the case of South East Asia, for example, Nambiar and Harwood (2014) identify considerable wood plantation potential, but this is dependent upon coordinated research and management for genetic improvement and soil protection.

By comparison, Australia and New Zealand have well developed institutional frameworks, research capacity and environmental policies for sustainable forest management and improved productivity. There has been over 100 years of plantation research and development focusing initially on exotic pines, and there is considerable scope to lift the productivity of the new eucalypt plantation estate in Australia. Genetic improvement and better site matching of species is expected to deliver higher growth for plantation eucalypts over the next rotation, given the rationalization of the former MIS estate in Australia. The New Zealand forest industry science and innovation strategy has also identified a 5 m^3 ha^{-1} $year^1$ increase in productivity by 2025.

The other key issue is what type of wood plantation industry is feasible or desirable in Australia and New Zealand into the future? This is a broader policy issue, reflecting the economics of exporting relatively unprocessed industrial wood versus domestic value adding, which will have implications for future national policy, plantation management and investment. Ferguson (2014a)argues that the Australian plantation industry is at a cross-roads and that a coherent national plantation policy is needed if the Australian government and industry wish to expand the plantation estate to maintain international competitive wood and paper manufacturing and associated rural employment.

Manufacturing costs in Australia and New Zealand can be at the higher end of global benchmarks, factoring in existing scale of facilities, environmental standards, labour and energy. For these reasons, there has been considerable interest in research and innovation into emerging opportunities to value add either low grade logs or under-utilised biomass residues such as for bioenergy, as well as higher value applications for sawn timber and engineered wood products.

Bioenergy offers potential as a new revenue stream that is dependent on renewable energy policy as well as available feedstocks and technology developments (International Energy Agency 2009). The Clean Energy Council (2008) identified there was enough wood waste from forestry and processing activity in Australia to generate up to 3000 Gwh per year or 7% of the national renewable energy target. In New Zealand, over $NZ13 million has been invested into research into converting forestry biomass into liquid biofuels by looking into the commercial viability of establishing a modular test plant to process forest waste into sustainable transport fuel (Ministry for Primary Industries 2013b). These opportunities reflect the growing demand for renewable energy sources and provide a medium to longer term market for the plantation wood industry.

However, it will be equally important that some of the social challenges of plantation development be managed into the future, given previous community concerns over plantation expansion and related socio-economic and environmental impacts. The need for improved community engagement is considered important in gaining wider acceptance from stakeholders (Gordon et al. 2013). Plantation development can occur across a continuum of planting types from monocultures of exotic species to stands of mixed native species; and varying scales from broad scale plantations to small farm woodlots and riparian plantings. The majority of wood plantations in Australia and New Zealand have been large-scale and restricted to a select number of species for high productivity and commercial purposes. Community preferences for more diversified and small-scale plantings may entail changes in planting design and species mix which can have direct impacts on the costs to the forest owner. Managing plantations as part of whole farm plans may assist in this process, although would need to be weighed against the commercial costs and benefits such as higher transaction costs from more dispersed plantation areas.

This leads to the other recent trend of increasing interest in planted forests for ecosystem services. To date, it has been difficult to quantify the extent of these types of plantings as they are undertaken by a range of individuals such as farmers, community based organisations, private businesses, family trusts and local governments. In Australia, there is estimated to be 153 000 ha of non-industrial planted forest.

The outlook for ecosystem services plantings other than for carbon sequestration in Australia and New Zealand is uncertain but likely to increase in conjunction with direct Government programmes, albeit on a small scale, for the amelioration of land degradation. In many cases these programmes are oriented toward revegetation using mixed plantings of endemic native species or natural regeneration. The other major activity is farm forestry activities for wood production, farm productivity and environmental amenity. In these cases, forestry extension is important and investment decisions can be more complex involving lifestyle, income, cash flow and risk choices for the landowner (Schirmer et al. 2000).

In terms of the broad potential for carbon plantings, either as not-for-harvest or joint carbon and wood plantations, it has been shown that prices above \$20/t CO_2 could stimulate new plantation investment in Australia and New Zealand. However, a number of significant impediments are identified in the respective national climate change policy schemes. In the case of Australia and New Zealand, this includes a lack of inclusion of harvested wood products in applicable methods and recognised afforestation activities and in Australia a relatively short contract period for the purchase of emission reductions by the Australian Government. In the case of New Zealand, it is also the partial coverage of the scheme for emitters and uncapped link to international carbon credits that has depressed carbon prices to less than \$3/t CO_2 in recent years. Other issues have included risk and uncertainty through volatility in carbon prices, particularly until underlying national and international carbon regulatory frameworks are fully developed.

CONCLUSIONS

Planted forest development in Australia and New Zealand has followed a similar pattern with a strong historical focus on industrial wood plantations. The net outcome from these policies has been the development of a sizeable wood plantation resource and associated downstream processing. However, once governments have transitioned away from direct incentives to an enabling policy environment there has been little new plantation investment.

Concurrently, markets have emerged for some ecosystem services such as landscape restoration, biodiversity and carbon sequestration. Further development and access to these markets may help address some of the investment hurdles for new wood plantations while delivering multiple benefits. There is therefore considerable interest in addressing market failure through the use of emerging markets for the joint production of wood and ecosystem services. A market based solution for ecosystem services in wood plantations has the added potential benefit of increasing economies of scale and competitiveness in wood resource availability and associated downstream processing.

It is argued that more coherent national plantation policies are needed to capture the full range of benefits from wood plantations and other planted forests. In 1997, the Plantations 2020 vision in Australia had a target of trebling the plantation area to 3 million hectares by 2020 to foster economic development. Similarly, the New Zealand government has a target of increasing the planted area by 20 000 hectares per annum, largely to meet climate policy objectives. These goals are unlikely to be met, at least in the short term, given falling rates of new plantings. In fact, there was a net decrease in Australia in the total area of wood plantations by 4 000 ha in 2011-12 (Gavran 2013). In New Zealand, recent surveys have identified deforestation intentions of many landowners, particularly conversion to dairy, over the next few years (Manley 2014). Better design of policy settings, particularly for carbon markets, and policy consistency between land uses, may curb falling rates of new plantings and encourage new plantings to help achieve climate change mitigation and economic development goals.

For existing wood plantations, growing demand in Asia, particularly in China and India, offers significant opportunities for export growth of round wood and woodchips from Australia and New Zealand. This outlook needs to be assessed against the growing supply of forest products from Latin America and Asia, the sustainability of which will depend on underlying policies for environmental management and institutional capacity. There is also potential for growth in bioenergy and other value added products from plantations in Australia and New Zealand over the medium term, either for domestic consumption or export.

In order to fully realise the opportunities for wood plantation development, enhanced community engagement will help address previous concerns regarding the acceptability of plantations in the rural landscape. Close attention will also need to be applied to managing future climate change risks, given forecast changes in temperature and rainfall and secondary effects such as increased pests and diseases or more extreme weather. Climate change forecasts present both risks and opportunities, through predicted increases or decreases in forest growth across regions and other risks posed by drought, fire and wind.

AUTHORS' CONTRIBUTIONS

David Rhodes and Michael Stephens jointly conceived the review as part of their participation in the Third International Congress on Planted Forests, contributing New Zealand and Australian planted forest perspectives respectively and underlying comparative analyses. Both authors reviewed the literature, drafted the thematic sections and approved the final manuscript.

ACKNOWLEDGEMENTS

The valuable suggestions made by two anonymous referees are gratefully acknowledged.

REFERENCES

1. Australian Bureau of Agricultural and Resource Economics and Sciences (2011). Potential effects of climate change on forests and forestry in Australia. Canberra, Australia: ABARES. August
2. Australian Bureau of Agricultural and Resource Economics and Sciences (2013). Preliminary long-term forecasts of wood product demand in Australia: research report 13.6. Canberra, Australia: ABARES. May
3. Australian Bureau of Agricultural and Resource Economics and Sciences (2014). Australian Forest and Wood Products Statistics: September and December quarters 2013. Retrieved 28 May 2014 from: http://www.daff.gov.au/ABARES/pages/publications/display.aspx?url=http://143.188.17.20/anrdl/DAFFService/display.php?fid=pb_afwpsd9abfe20140526_11a.xml
4. Australian Bureau of Statistics (2014). 3222.0 Population Projections, Australia, 2012 (base) to 2101. Retrieved 28

May 2014 from:http://www.abs.gov.au/ausstats/abs@.nsf/Lookup/3222.0main+features32012%20(base)%20to%202101

5. Asia Pacific Forestry Commission (2010). Asia-Pacific forests and forestry to 2020. Report of the second Asia-Pacific forestry sector outlook study. Bangkok: Food and Agriculture Organization of the United Nations.
6. Bahaus, J, van der Meer, P, Kanninen, M (2010). Ecosystem goods and services from plantation forests. London/Washington: Earthscan.
7. Baral, H, Keenan, RJ, Fox, JC, Stork, NE, Kasel, S (2013). Spatial assessment of ecosystem goods and services in complex production landscapes: a case study from south-eastern Australia. *Ecological Complexity, 13,* 35–45.
8. Battaglia, M, Bruce, J, Brack, C, Baker, T (2009). Climate change and Australia's plantation estate: analysis of vulnerability and preliminary investigation of adaptation options. (FWPA) Project No. PNC068-0709). Melbourne: Forest and Wood Products Australia/CSIRO.
9. Booth, TH, Kirschbaum, MUF, Battaglia, M In C. Stokes & M Howden (Ed.) (2010). Forestry.*Adapting agriculture to climate change, Chapter 9,* 137–152. Victoria, Australia: CSIRO Publishing.
10. Buongiorno, J, Zhu, S, Raunikar, R, Prestemon, JP (2012). Outlook to 2060 for world forests and forest industries: a technical document supporting the Forest Service 2010 RPA assessment. Asheville, North Carolina: United States Department of Agriculture, Southern Research Station.
11. Bureau of Rural Sciences (2005). Socioeconomic impacts of plantation forestry. Melbourne, Australia: Forest and Wood Products Research and Development Corporation. November
12. Carle, J, & Holgrem, P (2008). Wood from planted forests: a global outlook 2005-2030. *Forest Products Journal, 58,* 6–18.
13. Clean Energy Council (2008). Australian bioenergy roadmap: setting the direction for biomass in stationary energy to 2020 and beyond. Melbourne, Australia. September
14. Commonwealth of Australia (2013). Emissions Reduction Fund Green Paper. Canberra: Department of the Environment.

15. Cossalter, C, & Pye-Smith, C (2003). Fast-wood forestry: myths and realities. Forest Perspectives. Bogor: Center for International Forestry Research, Indonesia.
16. Dale, G, & Dieters, M (2007). Economic returns from environmental problems: breeding salt- and drought- tolerant eucalypts for salinity abatement and commercial forestry. *Ecological Engineering,31*, 175–182.
17. de Fegely, R, Stephens, M, Hansard, A (2011). Review of policies and investment models to support continued plantation investment in Australia. (Project No. PRA189-1011). Melbourne, Australia: Forest and Wood Products Australia. March
18. Dickinson, G, Reilly, D, Lindsay, DAJ (2011). Abstracts from Darwin 2011 - African Mahogany Plantation Industry Forum, Darwin, 30 August - 1 September 2011. Brisbane, Australia: Queensland Department of Employment, Economic Development and Innovation.
19. Fairweather, J, & Swaffield, S (2003). Public perceptions of natural character and implications for the forest sector. *New Zealand Journal of Forestry, 47*, 24–30.
20. Ferguson, I (2014a). Australian plantations: mixed signals ahead. *International Forestry Review,16*, 160–171.
21. Ferguson, I (2014b). Australian plantation inventory: ownership changes, availability and policy.*Australian Forestry, 77*, 25–38.
22. Food and Agriculture Organization (FAO) (2001). Mean annual volume increment of selected industrial forest plantation species. Forest Plantation Thematic Papers, Working Paper 1. Rome: Food and Agriculture Organization of the United Nations. April
23. Food and Agriculture Organization (FAO) (2006a). Global forest resource assessment 2005. (FAO Forestry Paper 147). Rome: Food and Agriculture Organization of the United Nations.
24. Food and Agriculture Organization (FAO) (2006b). Global planted forests thematic study: results and analysis. (Planted Forests and Trees Working Paper 38). Rome: Food and Agriculture Organization of the United Nations.
25. Food and Agriculture Organization (FAO) (2010). Global forest resources assessment 2010, main report. (FAO Forestry Paper

163). Rome: Food and Agriculture Organization of the United Nations.

26. Gavran, M, Frakes, I, Davey, S, Mahendrarajan, S (2012). Australia's plantation log supply 2010-2054. Canberra: Australian Bureau of Agricultural and Resource Economics and Sciences. May
27. Gavran, M (2013). Australian plantation statistics 2013 update. (Technical report 13.3). Canberra: Australian Bureau of Agricultural and Resource Economics and Sciences. May
28. Global Construction Perspectives and Oxford Economics (2013). Global construction market to reach $15 trillion by 2025. Retrieved 12 March 2014 from: http://www.globalconstruction2025.com/media/wysiwyg/press-release/GCP-2025-press-release.pdf
29. Gordon, M, Schirmer, J, Lockwood, M, Vanclay, F, Hanson, D (2013). Being good neighbours: current practices, barriers, and opportunities for community engagement in Australian plantation forestry. *Land Use Policy, 34*, 62–71.
30. Hale, T, Kahui, V, Farhat, D (2014). A modified production possibility frontier for efficient forestry management under the New Zealand Emissions Trading Scheme. *Australian Journal of Agricultural and Resource Economics, 58*, 1–17.
31. Hock, B, Payn, T, Clinton, P, Turner, J (2009). Towards green markets for New Zealand plantations. *New Zealand Journal of Forestry, 54*, 9–19.
32. Hennessy, K, Fitzharris, B, Bates, BC, Harvey, N, Howden, SM, Hughes, L, Salinger, J, Warrick, R In ML Parry, OF Canziani, JP Palutikof, PJ van der Linden & CE Hanson (Eds.) (2007). Australia and New Zealand. *Climate Change 2007: impacts, adaptation and vulnerability Contribution of Working Group II to the Fourth Assessment Report of the Intergovernmental Panel on Climate Change* (pp. 507–540). Cambridge, UK: Cambridge University Press.
33. Indufor (2012). Strategic review on the future of forest plantations. Retrieved 12 March 2014 from: http://ic.fsc.org/download.strategic-review-on-the-future-of-forest-plantations-full-report.672.htm
34. International Energy Agency (2009). Bioenergy - a sustainable and reliable energy source: a review of status and prospects.

(IEA Bioenergy: ExCo 2009:06, Main Report). Paris: International Energy Agency.

35. International Energy Agency (2011). Global wood pellet industry market and trade study. (IEA Bioenergy Task 40: sustainable international bioenergy trade). Paris: International Energy Agency. December

36. Kavanaugh, RP, Stanton, MA, Herring, MW (2007). Eucalypt plantings on farms benefit woodland birds in south eastern Australia. *Austral Ecology, 32*, 635–650.

37. Keenan, RJ, Davey, S, Grieve, A, Moran, B, Donaldson, J In HC Sim, S Appanah, & YC Youn (Eds.) (2004). Market mechanisms and assessment methods for environmental services from private forests in Australia. *Proceedings of the workshop Forests for poverty reduction: opportunities with Clean Development Mechanism, environmental services and biodiversity* Bangkok: Food and Agriculture Organization.

38. Kugelman, M, & Levenstein, SL (2012). The global farms race: land grabs, agricultural investment, and the scramble for food security. Washington DC: Island Press.

39. Lamers, P, Junginger, M, Hamelinck, C, Faaij, A (2012). Developments in international solid biofuel trade - an analysis of volumes, policies and market factors. *Renewable and Sustainable Energy Reviews, 16*, 3176–3199.

40. Langer, L, & Barnard, T (2002). Local community attitudes to plantation forestry, Gisborne/East Coast region. *New Zealand Journal of Forestry, 47*, 14–16.

41. Lindenmayer, DB, & Hobbs, RJ (2004). Fauna conservation in Australian plantation forests - a review. *Biological Conservation, 119*, 151–168.

42. Lippke, B, Oneil, E, Harrison, R, Skog, K, Gustavvson, L, Sathre, R (2011). Life cycle impacts of forest management and wood utilization on carbon mitigation: knowns and unknowns. *Carbon Management, 2*, 303–333.

43. Maclaren, JP (1996). Environmental effects of planted forests in New Zealand: the implications of continued afforestation of pasture. (Forest Research Institute Bulletin No. 198). Rotorua, New Zealand: New Zealand Forest Research Institute.

44. Malsheimer, RW, Bowyer, JL, Fried, JS, Gee, E, Izlar, R, Miner, RA, Munn, IA, Oneil, E, Stewart, WC (2011). Managing forests because carbon matters: integrating energy, products and land management policy. *Journal of Forestry, 109*(7S), S7–S50.
45. Manley, B, & Maclaren, P (2009). Modelling the impact of carbon trading legislation on New Zealand's plantation estate. *54*, 39–44. New Zealand Journal of Forestry.
46. Manley, B (2014). Deforestation Survey 2013 Final Report. Ministry for Primary Industries Technical Paper 2014/07, February.
47. McDermott, CL, Cashore, B, Kanowski, P (2010). Global environmental forest policies: an international comparison. (Chapter 8: Oceania: Australia and New Zealand). London, New York: Earthscan.
48. Ministry for Primary Industries (2013a). Situation and outlook for primary industries 2013, June. Retrieved 29 August 2014 from: http://www.mpi.govt.nz/news-resources/news/situation-amp-outlook-for-primary-industries
49. Ministry for Primary Industries (2013b). Stump to Pump programme receives PGP funding boost. [Media release, 25 July 2013]. Retrieved 10 June 2014 from: http://www.mpi.govt.nz/news-resources/news/stump-to-pump-programme-receives-pgp-funding-boost
50. Ministry for Primary Industries (2014). Forestry statistics. Retrieved 29 August 2014 from: http://www.mpi.govt.nz/forestry/statistics-forecasting/forestry-statistics
51. Mitchell, CD, Harper, RJ, Keenan, RJ (2012). Current status and future prospects for carbon forestry in Australia. *Australian Forestry, 75*, 200–212.
52. Montreal Process Implementation Group for Australia and National Forest Inventory Steering Committee (2013). Australia's State of the Forests Report 2013. Canberra: ABARES. December
53. Nambiar, S, & Ferguson, I (2005). New forests: wood production and environmental services. Melbourne, Australia: CSIRO Publishing.
54. Nambiar, EKS, & Harwood, CE (2014). Productivity of acacia and eucalypt plantations in South-east Asia. 1. Bio-physical

determinants of production: opportunities and challenges. *International Forestry Review, 16*, 225–248.

55. National Land and Water Resources Audit (2001). Australian dryland salinity assessment 2000: extent, impacts, processes, monitoring and management options. Canberra: National Land and Water Resources Audit.
56. New Zealand Forest Owners Association (2013). New Zealand plantation forest industry. Facts and figures. Retrieved 14 April 2014 from: http://www.nzfoa.org.nz/publications-5/facts-and-figures
57. Nielson, DA (2007). Corporate private sector dimensions in planted forest investments. (Planted Forests and Trees Working Paper Series FP/40E). Rome: Food and Agriculture Organization of the United Nations.
58. Oliver, CD, Nassar, NT, Lippke, BR, McCarter, JB (2014). Carbon, fossil fuel, and biodiversity mitigation with wood and forests. *Journal of Sustainable Forestry, 33*, 248–275.
59. Paul, KI, Reeson, A, Polglase, PJ, Ritson, P (2013). Economic and employment implications of a carbon market for industrial plantation forestry. *Land Use Policy, 30*, 528–540.
60. Pulp and Paper News (2014). Vietnam now world's largest woodchip supplier.Retrieved 9 June 2014 from: http://www.pulpapernews.com/2012/03/vietnam-now-world-s-largest-wood-chip-supplier
61. Rhodes, D (2001). Rehabilitation of deforested steep slopes on the east coast of New Zealand's North Island. *Unasylva, 207*, 21–29.
62. Rhodes, D, & Novis, J (2002). The impact of incentives on the development of plantation forest resources in New Zealand. (MAF Information Paper No. 45). Wellington, New Zealand: Ministry of Agriculture and Forestry.
63. Schirmer, J (2007). Plantations and social conflict: exploring the differences between small-scale and large-scale plantation forestry. *Small-scale Forestry, 6*, 19–33.
64. Schirmer, J, Kanowski, P, Race, D (2000). Factors affecting adoption of plantation forestry on farms: implications for farm forestry in Australia. *Australian Forestry, 63*, 44–51.

65. Singh, S, Davey, S, Cole, M (2010). Implications of climate change for forests, vegetation and carbon in Australia. *New Zealand Journal of Forestry Science, 40*, 141–152.

66. Sochacki, SJ, Harper, RJ, Smettem, KRJ (2012). Bio-mitigation of carbon following afforestation of abandoned salinized farmland. *Global Change Biology (GCB) Bioenergy, 4*, 193–201.

67. State of the Environment Committee (2011). Australia state of the environment 2011. (Independent report to the Australian Government Minister for Sustainability, Environment, Water, Population and Communities). Canberra.

68. Stephens, M (2010). Bushfire, forests and land management policy under a changing climate. *Farm Policy Journal, 7*, 11–19.

69. Stephens, M, Pinkard, L, Keenan, R (2012). Plantation forest industry climate change adaptation handbook. Canberra: Australian Forest Products Association. [http://www.ausfpa.com.au/wp-content/uploads/Change-Adaptation-Handbook.pdf]

70. Stephens, ML, & Grist, P (2014). Market failure for plantations: past experiences and emerging trends for delivering wood production and ecosystem services in Australia. *International Forestry Review, 16*, 205–215.

71. Townsend, PV, Harper, RJ, Brennan, PD, Dean, C, Wu, S, Smettem, KRJ, Cook, SE (2012). Multiple environmental services as an opportunity for watershed restoration. *Forest Policy and Economics,17*, 45–58.

72. URS Forestry (2008). Farm forestry area and resources in Australia. (RIRDC Publication No. 08/104). Canberra: Rural Industries Research and Development Corporation.

73. Watt, MS, Kirschbaum, MUF, Paul, TSH, Tait, A, Pearce, HG, Brockerhoff, EG, Moore, JR, Bulman, LS, Kriticos, DJ (2008). The effect of climate change on New Zealand's planted forests: impacts, risks and opportunities. (Ministry of Agriculture and Forestry Report No. CC MAF POL_2008-07 (106-1) No. 1, July). Wellington, New Zealand.

74. West, G, Wakelin, S, Wall, A, Turner, J, Poole, B (2011). Achieving multiple regional goals with carbon forestry. (A Scion report for Waikato Regional Council). Rotorua, New Zealand: Scion. [http://www.waikatoregion.govt.nz/PageFiles/19232/scion%20 report.pdf]

75. Williams, K (2008). Community attitudes to plantation forestry. (Technical Report 194). Cooperative Research Centre for Forestry and Department of Resource Management and Geography, University of Melbourne.

76. Wilson, S, Bhati, UN, Stephens, M (1995). Trees on Australian broadacre and dairy farms. Quarterly forest products statistics: September quarter. Canberra: Australian Bureau of Agricultural and Resource Economics.

77. Wood Council of New Zealand (2012). New Zealand forest and wood products industry strategic action plan. Wellington, New Zealand. Retrieved 12 March 2014 from:http://www.woodco.org.nz/images/stories/pdfs/ForestWood_Strategic_Action_Plan.pdf

78. Yao, RT, Scarpa, R, Turner, JA, Barnard, TD, Rose, JM, Palma, JHN, Harrison, DR (2014). Valuing biodiversity enhancement in New Zealand's planted forests: socioeconomic and spatial determinants of willingness-to-pay. *Ecological Economics, 98,* 90–101.

79. Zhang, L, Dowling, T, Hocking, M, Morris, J, Adams, G, Hickel, K, Best, A, Vertessy, R (2003). Predicting the effects of large-scale afforestation on annual flow regime and water allocation: an example for the Goulburn-Broken catchments. (Technical Report 03/5). Canberra: CRC for Catchment Hydrology.

80. Zhang, L, Vertessy, R, Walker, G, Gilfedder, M, Hairsine, P (2007). Afforestation in a catchment context: understanding the impacts on water yield and salinity. (Industry report 1/07). Melbourne, Australia: eWater CRC.

Chapter 7

Overview of Principles and Implementations to Deal with Spatial Issues in Monitoring Environmental Effects of Genetically Modified Organisms

Winfried Schröder and Gunther Schmidt

Chair of Landscape Ecology, Universität Vechta, Vechta, 49364, Germany

ABSTRACT

The approval of genetically modified organisms [GMO] for deliberate release and placing on the market requires GMO environmental risk assessment [ERA] and GMO environmental monitoring [EM]. Both GMO ERA and GMO EM are still under discussion. The goal of this article is, firstly, to analyse principles of CMO EM as published in the Association of German Engineers [VDI] Guideline 4330 Part 1, focusing on the characterisation of the receiving environment affected by GMO cultivation and the representativeness of GMO EM to assess large-scale implications of GMO cultivation. Secondly, the article introduces measures to meet these issues by the use of map data and statistics within a geographical information system [GIS]. Finally, three case studies exemplify the application of data and methods. To deal with spatial issues of GMO EM as outlined in the VDI Guideline 4330 Part 1, a GIS-based approach is presented. It relies on both spatial data collected from several sources which were derived from sample point data and geostatistical and multivariate statistical methods within a GIS environment. Data used for describing the receiving environment and for planning and evaluating monitoring schemes comprise information about land use, climate, phenology, soil coverage, species distribution and ecoregions. The case studies deal with (1) ecological land classification for characterisation of GMO-receiving environments and representative EM, (2) selection of representative sites for modelling GMO dispersal, and (3) delineation and mapping of segregation distances. Even a systematic and stepwise-structured risk assessment cannot cover all risk relevant questions, especially large-scale, long-term and combinatory effects which may not occur before the conventional application of the respective GMO. Hence, GMO EM is crucial to deal with unanticipated and undesirable effects. The article gives an overview of a GIS implementation and relevant geodata promoting GMO EM.

BACKGROUND

In the European Union [EU], the release of genetically modified organisms [GMO] into the environment is regulated by EU Directive 2001/18/EC [1]. Accordingly, post-market environmental monitoring of genetically modified plants [GMP EM] has to be implemented to detect and prevent adverse effects on human health and the environment. However, no general strategies for GMP EM have been established so far. In Germany, one EM strategy discussed is the Guideline 4330 Part 1 published by the Association of German Engineers [VDI] [2]. It applies to the monitoring of ecological effects of GMP, but does not address possible effects of GMP on human health. Contrary to the directive of the European Community [1] and the study of Sanvido et al. [3], the guideline of the VDI [2] does not differentiate between case-specific monitoring [CSM] and general surveillance [GS]. CSM should focus on anticipated effects of a specific GMP based on pre-market risk assessment, whereas GS is designed to detect unanticipated adverse effects which were not covered by risk assessment comprising, for instance, cumulative and long-term effects.

The VDI [2] covers *ecological effects* of GMP encompassing direct, indirect, immediate and delayed as well as cumulative long-term effects. Environmental effects of GMP can occur on several *levels of ecological organization* in terms of structures and underlying functions, and correlated *levels of time and space* which have to be covered by GMP EM [4]. The VDI [2] provides planning and implementation criteria for GMP EM and forms the framework for technical instructions with respect to the levels mentioned above, *items to be protected, protection targets*and *checkpoints*. GMP EM should allow evaluating the condition of the items to be protected and to track the accomplishment of protection targets. The required parameters have to be collected using validated and standardised methods.

Since protection targets and checkpoints may not only be influenced by GMP, it is necessary to differentiate between GMP-related effects and those that are not related to GMP. Accordingly, data on items to be protected and protection targets without influence of GMP must also be compiled. To reach this requirement, temporal and spatial comparisons are needed: The environmental baseline status prior to the introduction of GMP has to be compared to the situation after GMP release regarding the selected checkpoints. The reliability of reference data depends on the period within the reference conditions were monitored before introducing GMP *(temporal comparison)*. Additionally, GMP-free regions have to be compared to those where GMP are cultivated. This requires a reference system where both conditions are monitored simultaneously *(spatial comparison)*. Reference sites should be as similar as possible to GMP-influenced areas considering the receiving agricultural environment. GMP areas and reference areas, however, can also be subjected to changes that are not caused by GMP. Thus, a thorough selection of monitoring areas is essential for monitoring potential ecological effects due to GMP cultivation. Besides the GMP fields and their surroundings, representative types of ecosystems that will potentially be affected should be considered as well. EM areas should be selected using a statistically substantiated procedure according to technically suitable representation criteria. The EM areas should be linked to other appropriate EM networks. In the long term, spatial rearrangement of EM areas is necessary regarding new effect relationships and spatial arrangement of land use patterns.

The GMP EM measuring data should be analysed on the basis of metadata describing them and by suitable (geo)statistical procedures. The documentation of measured variables, methods, survey intervals and areas must be carried out according to standard methods and using a main meta-database or several interrelated databases. Meta-databases should help evaluate on to what extent the data records can be compared with one another for assessment.

Based on the basic considerations as laid down in the guideline of the VDI [2] and summarised above, some research projects aimed at dealing with GMP EM at a landscape level and at developing techniques for supporting the application of the respective EM strategies. In the following, we refer to some of the respective methods and results and, thereby, concentrate on the setting in Germany as an example.

METHODS AND DATA

The following sections contain an overview of procedures implemented in a *geographic information system* [*GIS*] including *geostatistics, multivariate statistics* and *geodata* to (1) characterize the GMP-receiving environment, (2) to assess the spatial representativeness of GMP EM sites and (3) to assess large-scale and long-term effects of GMP cultivation. In Germany, several research projects dealt with these issues, and some of the methods applied and results achieved are outlined. It is shown that geodata are useful to describe the receiving environment in the near and far vicinities of GMP fields. Statistical analyses and classification of geodata are presented which serve to derive ecoregions, e.g. climatic and agricultural patterns and, thereby, help for assessing the representativeness of running or planned GMP EM sites and for investigating adverse ecological effects of GMP release on different spatial scales and for different agricultural regimes [5-11].

Geostatistics is a point-pattern analysis that generates surface predictions from data points. This relies on investigating and modelling the spatial autocorrelation among sample data by variogram analysis.

In order to apply kriging for interpolation, it is necessary to adapt a defined variogram model to the experimental variogram. Based on the variogram model, several kriging methods can be used for spatial predictions which finally are mapped [12]. For the interpretation of the kriging estimations, a cross-validation has to be performed.

Multivariate statistics such as cluster analysis or tree-based models, two of them are the classification and regression trees [CART] and chi-squared automatic interaction detection (CHAID), serve to spatially differentiate the multiple relationships between geodata stored in a GIS. Based on these relations, predictions in time and space become possible as well as the characterisation of the receiving environment in terms of ecoregions [13-18]. In the context of GMP dispersal, *cluster analysis* can be used to integrate measurement data from different meteorological networks with different coverage in a GIS environment for defining representative climatic regions. Climatic regions together with an ecological land classification were used to stratify the receiving environment in order to select a representative number of sites for modelling the GMP dispersal [10].

Geodata on meteorology, land use, local biodiversity and agricultural management schemes are needed for monitoring and modelling dispersal and persistence of GMP as well as planning GMP EM with respect to coexistence issues in agricultural landscapes. The data described in the following have been collected from several sources or have been calculated from sample point data by the use of the above mentioned statistical methods in a GIS environment described by Kleppin et al.[19].

Land use data can be obtained from either satellite images, GIS data collected during field experiments, cadastral surveys provided by local land registries, vertical air photographs or the Common Agricultural Policy notifications, each type of source being used at different scales and consequently provide different spatial and semantic resolutions. To some extent, data on field geometries providing detailed information on agricultural land use can be obtained from the Integrated Administration and Control System (InVeKoS) database, which is an important tool for the EU member states to regulate agricultural subsidies. In fact, due to legal restrictions and inconsistencies in data harmonisation which is due to federal responsibilities, this dataset is not available for public use [9]. Based on satellite images, data on land use patterns are offered by the European Topic Centre on Land Use and Spatial Information [20], where the distribution of the CORINE Landcover maps is administrated [21]. Data on the cultivation of crops in Europe are available at the Statistical Office of the European Communities [22], which offers data on various topics, among of which is also agriculture. The main cropping areas of oilseed rape are located in northeast Germany as well as in the Alsace in France. In these regions, oilseed rape is cultivated on up to 25% of the arable land. Due to the increased cultivation of energy plants, it can be assumed that the cultivation of maize (biogas) and oilseed rape (biodiesel) will be intensified in the future. For Germany, it can be stated that in 2007, there was an increase in maize cultivation of 9.6% and of 8.8% for oilseed rape cultivation compared to those in 2006.

For large-scale analyses of GMP impacts, *meteorological data* are needed. These are, for example, data on precipitation, air temperature, sunshine duration, the number of frost days and wind conditions. Climate affects the growth, persistence and dispersal of GM crops and their pollen and seeds. These data could be retrieved from meteorological stations, which are usually widespread in Europe. However, depending on the required climatic element (precipitation, air temperature, wind or solar radiation), the number of monitoring sites and, thus, the validity of assumptions based on these data are different. For example, in France, the spatial density of monitoring sites collecting data on precipitation is two times more dense than on temperature, four times more than on wind and ten times more than on solar radiation. In Germany, the number of meteorological monitoring sites differs quite more. The German Weather Service operates about 4,400 precipitation sites, but only 660 stations for air temperature and 220 for solar radiation. Therefore, interpolations or extrapolations may be necessary, covering the whole territory of a country. For Europe, free datasets with a resolution of 10 arc min (approximately 20 × 20 km) are available at the Climatic Research Unit (CRU) [23,24]. For modelling the pollen transport, phenological data on the flowering of GM crops should be considered, too. It should be taken into account that global warming might have changed the temperature-induced beginning of rape and maize bloom [18,25]. Furthermore, modelling pollen dispersal requires data on wind regimes. The dynamics of pollen transport can be described by compiling and processing data on wind direction and velocity. The wind direction influences the transport direction of the pollen and, thus, potential areas of exposure. Given a constant emission rate, the wind velocity affects the range and the transport speed of airborne pollen and leads to a dilution (stretching); as with higher wind velocities, a larger air volume passes the source surface, and the concentration per unit volume is reduced [26].

Data on soil texture and soil types are available from the Food and Agriculture Organization [FAO]: (1) the Digital Soil Map of the World (about 10 × 10 km^2) [27] and (2) the Harmonized World Soil Database (about 1 × 1 km^2) [28]. Data on the potential natural vegetation which can be used for ecological land classification can be obtained from the Federal Agency of Nature Conservation (BfN) in Germany [29]. The potential natural vegetation [PNV] map stratifies Europe into more than 700 PNV units. The PNV can be defined as the vegetation that could be established without human interference under present climatic and soil conditions and is an integral indicator for the ecological conditions in terrestrial ecosystems [16].

For *biodiversity data* in the detection of adverse effects on biodiversity, a link between GMP and biodiversity monitoring is imperative [30,31]. It has to be expected that due to a large-scale commercial use of GMP, adverse effects on biodiversity become substantial. Biodiversity monitoring schemes could provide information on potential threats induced by GMP. For instance, biodiversity monitoring is able to detect the potential invasiveness of GM crops and the potentially enhanced mortality of non-target organisms, and it may also draw a more general picture on potential effects on the countryside biodiversity. In Europe, several biodiversity monitoring networks exist due to the Convention on Biodiversity, which commits its signatory countries to identify and monitor national biodiversity. However, these monitoring networks are poorly connected, and data are usually available only on a local or national level [32], whereas the monitoring of birds and butterflies is well established over long periods in some European countries (e.g. > 30 years in the UK), allowing an assessment of changes at several trophic and geographical scales [33,34]; monitoring is not in the same quality established across taxonomic groups relevant for GMP EM. Only few larger scale monitoring schemes of plants exist [35]. As of September 2007, the EuMon database comprised 552 complete monitoring schemes covering approximately 4,000 species and 145 different habitat types and addresses of 239 monitoring coordinators and institutions. Furthermore, the database contains information on sampling methods.

Changes of biodiversity due to GMP cultivation must be extractable from the background noise of sampling variability and population fluctuations. This is only possible if a considerable amount of sites is frequently and accurately monitored and if reference areas, i.e. areas without potential influence of GMP, are monitored at the same time and with the same accuracy. Even though the EuMon database is the largest collection of metadata on biodiversity monitoring available, it is not comprehensive and might be confounded by biases in observation accuracy [36]. Besides the EuMon database, there are only few more data sources where information on biodiversity or distribution of plant species - that may, for instance, serve as crossbreeding partners of GMP - may be obtained. The Global Biodiversity Information Facility [37] enables free and open access to biodiversity data worldwide via the Internet to support sustainable development. An information system was built to allow the linkage of diverse data types from disparate sources, promoting capacity building and catalysing development of analytical tools for improved decision-making. A special application concerning forest data and the distribution of forest tree species is available through the European Forest Genetic Resources Programme [38], which is a collaborative programme among European countries to promote conservation and sustainable use of forest genetic resources. There is information available describing the spatial distribution of about 40 tree species occurring all over Europe. Data are stored as JPEG files but also as shape files for usage within a GIS environment. DIVA-GIS [39] is a free and open-source GIS to generate and analyse worldwide maps on species distribution data. DIVA-GIS was developed at the International Potato Center [40]. In Germany, the Federal Agency for Nature Conservation (BfN) maintains the web application FloraWeb [41], where information on about 3,500 plant species are stored containing details on e.g. taxonomy, biology and spatial distribution of plants in Germany. An interactive web application illustrates the distribution of the PNV [29] in Germany. A Java applet allows mapping selected plant species in a spatial differentiation based on cadastre maps (scale 1 is 25,000; $\approx 11 \times 11$ km^2).

A crucial problem for spatial analyses is the availability of *data on the distribution of present pests*. For the federal state of Brandenburg, there were data collected on a district level regarding the spatial distribution of the European corn borer (*Ostrinia nubilalis*) which was one reason for the introduction of Bt maize. Figure 1 depicts the distribution of the corn borer in the federal state of Brandenburg for the years 2005, 2006 and 2007.

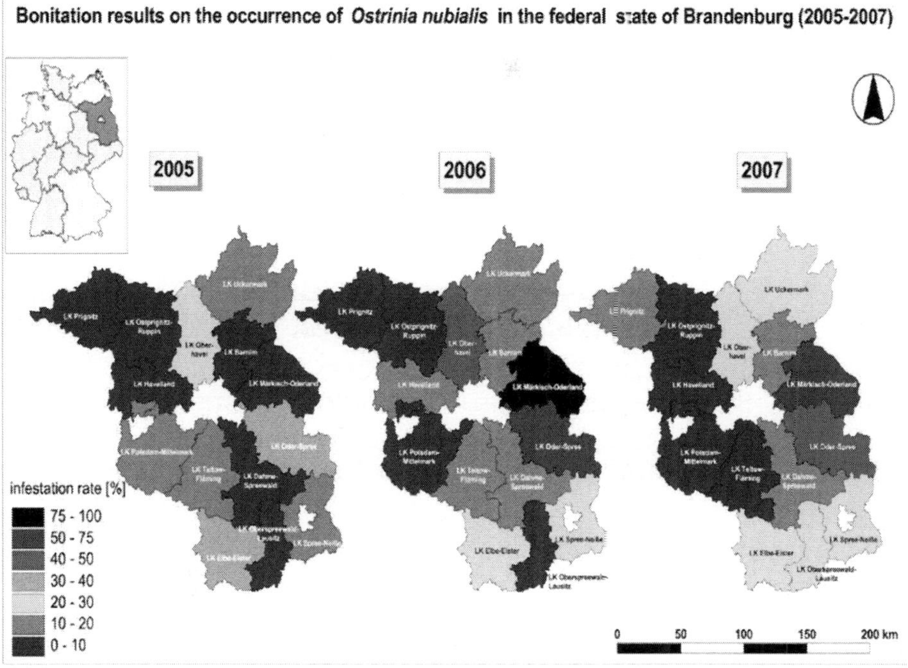

Figure 1: Distribution of the corn borer (*O. nubilalis*) in Brandenburg between 2005 and 2007. Mapping is based on data handed over by Dr. Werner Kratz, Landesumweltamt Brandenburg.

Some of the data and methods presented above have been used in several case studies which have been conducted according to selected aspects of the VDI Guideline 4330 Part 1 [2]. Three of them are summarised below: (1) ecological land classification for characterisation of GMP-receiving environments and representative EM, (2) selection of representative sites for modelling GMP dispersal and (3) delineation and mapping of isolation zones.

RESULTS AND DISCUSSION

Case Study 1: Ecological Land Classification for Characterisation of the GMP-Receiving Environment and Implementation Representative EM

The VDI Guideline 4330 Part 1 [2], the German Federal Nature Protection Law (Section 6 of the Bundesnaturschutzgesetz), the environmental monitoring concept of the Federal Ministry for the Environment, Nature Conservation and Nuclear Safety [42] as well as the preamble of the administrative agreement between the German government and the federal states on the exchange of environmental data specify the following targets that should be complied with when carrying out environmental monitoring: The monitoring should be coordinated and based on harmonised or standardised methods [43] so that the data can be compared and used for statistical analysis and modelling. The monitoring data should allow for spatial extrapolation in order to bridge geographical gaps and for supporting long-term research on environmental changes. The flow of data should be efficient, and the data should be available for scientists, especially for statistical testing of hypotheses and modelling data. The latter aspect also implies important technical issues because of the enormous amount of information and data collected. For example, environmental monitoring networks require information exchange, which has to be supported by an adequate and efficient information platform that handles documentation and exchange of metadata (site descriptions, quality control data), measuring data and geodata. An appropriate tool to achieve these goals is the implementation of a web-based GIS that contains relevant geodata and offers tools for integration of information of related environmental monitoring networks and tools for data analysis. Such information platforms help in reducing GMP EM costs and enable data and information exchange between different stakeholders, involving farmers, legal authorities and the public.

Long-term and indirect effects of any new technology present a challenge to risk assessment. Post-release GMP EM provides mechanisms for the early detection of any adverse effects, but the challenge for scientific committees, applicants and regulators is to identify the key areas of uncertainty and to design appropriate monitoring and surveillance methods. Small plots and laboratory studies are unlikely to prove useful in such an evaluation. Therefore, appropriate large-scale monitoring, experimentation and modelling are needed to determine the impact on the landscape from GMP trait characteristics [44]. GMP monitoring should cover both the GMP concerned and the potential receiving environment. GMP ERA and EM should comprise the evaluation of the characteristics of the GMP and its effects and stability in the environment, combined with ecological characteristics of the environment in which the introduction will take place. Thus, EM of GMP impacts should be implemented regarding description, explanation and modelling of environmental changes potentially due to GMP cultivation.

The requirements mentioned above imply that the EM network should cover the ecologically defined land classes in the respective country without gaps by a statistically adequate number of EM sites. This ecological representativeness is crucial for the validity of the EM sampling data[45,46]. Thus, monitoring and modelling of GMP dispersal should be performed at locations which are representative for larger areas with respect to those factors which potentially influence the dispersal, as for instance natural land characteristics such as wind conditions. Following this concept, ecoregions can be used to extrapolate modelling results (up-scaling) calculated for specific agricultural and environmental conditions at single locations to those areas where similar conditions exist, i.e. regions belonging to the same ecoregion. Additionally, GMP EM should take place in areas exposed to GMP, preferably cultivated fields and their environment, but should also include regions with no or unknown GMP exposure as reference areas. On a case-by-case basis, depending on the GMP characteristics, the selected indicators, checkpoints and related analytical methods should consider different relevant spatial and temporal scales [2,6]. The number of monitoring sites and regions needs to be sufficient to support statistical analysis of results based on good scientific practice [47-49]. For each GMP monitoring, design and data analyses should be based on appropriate scales of space and time, and the quality and quantity of data should be representative and interpretable. Criteria for selecting monitoring sites and regions include representativeness of sites cultivated with specific GMP, with emphasis on regions repeatedly cultivated with GMP; representativeness of ecological regions containing the spectrum of relevant indicators; availability of sites already monitored within other environmental programmes; and areas with environmental conditions facilitating spread or survival of GMP [4,50].

In order to check the representativeness of existing EM networks which might be appropriate for EM GMP or for establishing specific EM GMP networks, ecoregionalisations are appropriate measures. For Europe and Germany, ecological land classifications were calculated by means of multivariate statistics and based on digital maps depicting the spatial patterns of ecologically relevant land characteristics. For both Germany and some federal states, ecoregions were calculated by applying CART and using surface maps on climate, altitude, soil and potential natural vegetation [6,16]. The resulting maps have a spatial resolution of 2 × 2 and 1 × 1 km^2. The land classification calculated for Europe by means of CART [13] subdivides the whole territory into ecoregions mapped in a grid with a cell size of about 20 × 20 km^2. Data used for calculating the ecoregions are maps on the PNV [29], on altitude (Global Land One-kilometer Base Elevation/GLOBE) [51], on soil texture (Digital Soil Map of the World/DSMW) [27] as well as on monthly averages on air temperature, sunshine duration, relative humidity and precipitation (Global Climate Dataset CL 2.0) [23]. The PNV was set as the target variable, whereas the above mentioned maps on altitude, soil texture and climate were chosen as predicting variables. In order to obtain a concise amount of ecoregions, the most detailed map depicting the spatial pattern of about 200 ecoregions was reduced to 40 ecoregions (Figure 2). Each of them can be described statistically and by the use of annual course diagrams and histograms as it is demonstrated for selected ecoregions (D_7 to D_22) in Figure 3.

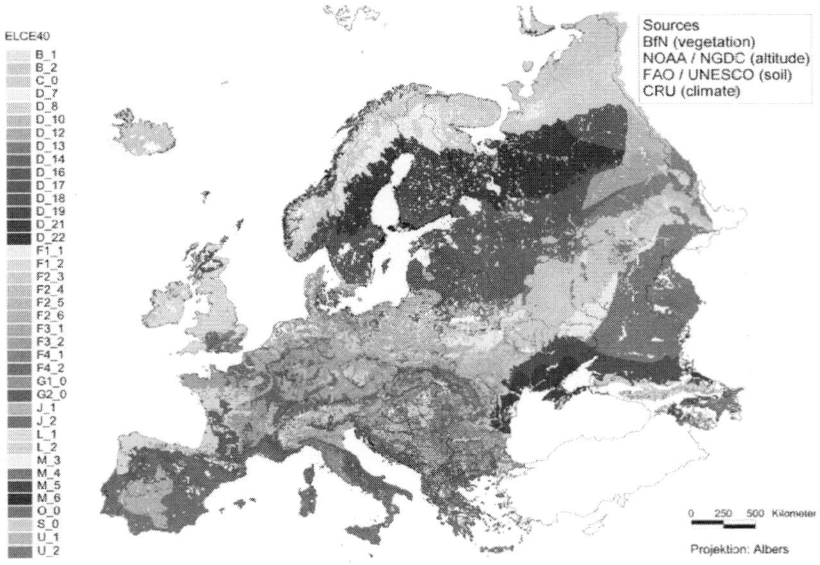

Figure 2: Ecoregions of Europe calculated by CART[13].

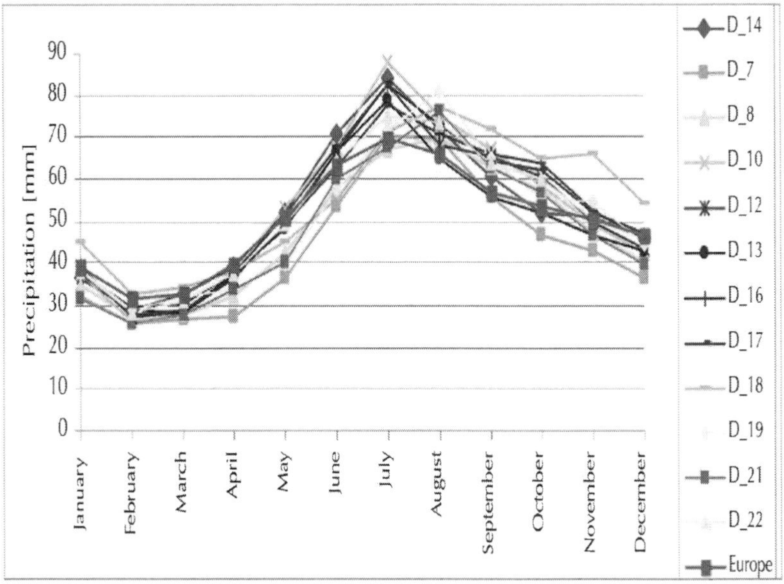

Figure 3: Annual course of precipitation (monthly means, 1961 to 1990) for some ecoregions in Europe [13].

Case Study 2: Selection of Representative Sites for Modelling GMP Dispersal

For modelling pollen dispersal of genetically modified oilseed rape [GM OSR], representative locations should be determined [5,10]. Accordingly, a method was developed that includes both the determination of representative OSR locations for modelling the dispersal at a field scale and the subsequent generalisation of the modelling result to the landscape level at a regional scale (up-scaling). Accordingly, land characteristics which are relevant for dispersal and persistence of GM OSR were regionalised within a GIS environment. The beginning of flowering of OSR was mapped by means of geostatistics. The resulting maps were used to select satellite images for the detection of OSR fields and to determine the period for the individual-based modelling. The monthly means (1961 to 1990) of precipitation [P], air temperature [T] and sunshine duration [S] were regionalised by the Ward cluster analysis [52], which has a wide range of applications in landscape ecology [53-55]. The PTS clusters were combined to four climatic regions which, together with Ward clusters on wind speed and direction as well as with land use clusters (crop rotation and management) [56], enabled to define eight regions in Northern Germany with a maximum of internal homogeneity. A distinct meteorological station was selected to represent each of these regions. Data on wind speed and direction (hourly means), precipitation, sunshine and air temperature (daily) measured at that location were provided for modelling the growth, dispersal and persistence of GM OSR on selected fields on the local level [57]. Linking each of the modelled sites with a map on German ecoregions [16], which integrates the spatial patterns of soils, elevation, vegetation and climate, the modelling results were anticipated by analogy reasoning to be valid for all those ecoregions which are represented by the modelling sites and, thus, could be spatially generalised for up-scaling [58].

Case study 3: Delineation and Mapping of Isolation Zones

Concerning the protection of non-target organisms that might be harmed due to GMP cultivation, a methodology was developed to classify the susceptibility/sensitivity of nature reserves [NSG] in Germany as being part of the receiving environment that might be affected due to GMP cropping in their vicinity. Within the joint research project 'Recommendations for isolation distances concerning the cultivation of genetically modified plants in the neighbourhood of protected areas' funded by the Federal Agency for Nature Conservation (BfN), possible risks for biocoenoses in protected areas were evaluated as well as measures which could mitigate or hinder negative effects [7]. According to Section 23 of the German Federal Nature Protection Law [BNatSchG], NSG are to protect nature and landscape properties by preserving and developing as well as by re-establishing existing biotopes of wild and endangered species. According to Section 34a of the BNatSchG, the use of GMP has to be accompanied by an environmental impact analysis of possible risks like it has to be done in projects affecting the integrity of Flora-Fauna-Habitats (FFH) or European bird sanctuaries. In order to classify NSG according to their potential endangerment by GMP invasion, a methodology based on GIS techniques and statistical measures was developed. Additionally, it examined what implications would emerge when introducing different isolation distances concerning the cultivation of herbicide-resistant OSR and insect-resistant maize near protection areas [9]. Both should help in monitoring and modelling GMP impacts. Within a GIS environment, geometries of conservation areas, land use data (CORINE Landcover) [21], agricultural information on the district level (Easystat: Statistik Regional 1999) as well as a map of German ecoregions [16] were integrated. All NSG were classified with respect to geometric properties and different intensities of cultivation area in their vicinity. The classification was realised by calculating a geometric coefficient [GC] which described the ratio of the buffer zone and the NSG area in order to parameterize the risk of GMP invasion. The smaller and/or the narrower the NSG, the larger is the buffer zone, relatively, and the higher is the risk for GMP invasion. According to frequency analyses of the GC, three percentile classes (low/medium/high) were derived. The cultivation area of maize and OSR cropping in the buffer zone around the NSG was expressed by a cultivation coefficient [CC]. This was calculated by adding up the area of maize and OSR cropland within a radius of 800 m (maize) and 4,000 m (OSR) around the NSG. Considering GMP cultivation in the future, these GMP fields are likely to be located in those regions where cultivation of conventional crops already has taken place. On the other hand, conventional maize or OSR fields might act as stepping stones to establish transgenes from GMP fields far off by cross-breeding with conventional stands, volunteers or ferals. Again, three percentile classes were built by frequency analyses. They describe the spectrum from a low to a high cultivation intensity of maize or OSR in the neighbourhood of each of the 7,338 NSG in Germany. The combination of GC and CC resulted in a total of nine risk categories [RC], describing the potential risk of endangerment by GMP cultivation in the vicinity of NSG. Areas with the highest risk were grouped in RC 9: Here, those NSG were assembled showing the smallest acreage and the highest cultivation rate of the respective crop (maize, OSR) in the neighbourhood of the NSG. With a numerical proportion of 7%, those sites cover only 0.4% of the total area of all NSG. All NSG showing the highest CC values had a total proportion of 60% [9].

CONCLUSIONS

The GMP EM is an important element of the regulatory framework for GMO cultivation in Europe and needs to be conducted according to scientifically sound methods and quality criteria to generate data which have to be robust and conclusive. The choice of parameters, methods and experimental designs of the locations and the timeframe for GMP EM needs to ensure that adverse effects of GMP and their use can be detected reliably and as early as possible. To reach this end, guidelines such as that of the VDI [2] are needed in attempting to harmonise and standardise the GMP EM design.

The VDI [2] recognizes that the environmental effects of GMP may vary with the characteristics of different receiving environments in terms of e.g. climate, soils, land use patterns or geographic distribution of wild relatives of certain GMP. Therefore, data derived by ERA or EM should be collected in those regions which are representative for respective ecological and agronomic characteristics which potentially could influence the spread and impacts of GMP. Thus, spatially differentiated monitoring schemes are needed, in particular with regard to biodiversity (e.g. non-target organisms) and ecological processes and functions (e.g. soil functions) in which these organisms are involved. However, access to relevant geodata is a prevalent problem.

In this context, the EU directive Infrastructure for Spatial Information in Europe [INSPIRE[a]] is an ambitious initiative to promote standardised data retrieval. In Germany, PortalU[b] is a first step to achieve the INSPIRE goals. However, the problem so far is that only few geodata sets are available, less of them being appropriate for GMP EM use. Exposure assessment is crucial for GMP EM, aiming to assess whether relevant parameters, e.g. certain non-target species, have to be in focus in the course of the monitoring. In combination with an effect assessment, the exposure assessment allows the evaluation of species which may be at risk. Geodata, ecological land classification, spatial estimation and GIS techniques in combination with dynamic modelling are fundamental to address effects on a landscape scale and long-term implications, to analyse and evaluate the appropriateness of existing monitoring programs or data for GMP EM, to design adaptations or extensions of the scope of GMP EM if they are inappropriate and to address the specific requirements for GMP EM.

Footnote

[a]http://www.ec-gis.org/inspire

[b]http://portalu.de

AUTHORS' CONTRIBUTIONS

GS performed the GIS and statistical analysis. WS conceived the study, participated in its design and coordination, and drafted the manuscript. Both authors read and approved the final manuscript.

REFERENCES

1. European Community (EC): *Directive 2001/18/EC of the European Parliament and of the Council of 12 March 2001 on the Deliberate Release into the Environment of Genetically Modified Organisms and Repealing Council Directive 90/220/EEC*. Brussels; 2001.
2. Association of German Engineers (VDI): *Monitoring the Effects of Genetically Modified Organisms-Genetically Modified Plants.*

Basic Principles and Strategies. VDI 4330 Part 1. Düsseldorf; 2006.

3. Sanvido O, Widmer F, Winzeler M, Bigler F: A conceptual framework for the design of environmental postmarket monitoring of genetically modified plants *Environ Biosafety Res* 2005, 4:13-27.
4. Züghart W, Breckling B: Konzeptionelle Entwicklung eines Monitoring von Umweltwirkungen transgener Kulturpflanzen, Teil 1 und 2 *UBA-Texte* 2003, 50/03:1-543.
5. Breckling B, Reuter H, Middelhoff U, Glemnitz M, Wurbs A, Schmidt G, Schröder W, Windhorst W: Risk indication of genetically modified organisms (GMO): modelling environmental exposure and dispersal across different scales: oilseed rape in Northern Germany as an integrated case study *Ecol Ind* 2011, 11:936-941.
6. Graef F, Schmidt G, Schröder W, Stachow U: Determining ecoregions for environmental and GMO monitoring networks *Enviro Monit Assess* 2005, 108:189-203.
7. Menzel G, Lünsmann I, Middelhoff U, Breckling B, Schmidt G, Tillmann J, Schröder W, Filser J, Reuter H: Abstandsregelungen beim Anbau gentechnisch veränderter Pflanzen (GVP) in der Nähe von Schutzgebieten. *Naturschutz und Biologische Vielfalt* 2005, 10:1-227.
8. Reuter H, Middelhoff U, Graef F, Verhoeven R, Batz T, Weis M, Schmidt G, Schröder W, Breckling B: Information system for monitoring environmental impacts of genetically modified organisms. *Environ Sci Pollut Res* 2010, 17:1479-1490.
9. Schmidt G, Schröder W: GIS analyses on possible risks for nature reserves due to genetically modified crops. In *Geospatial Crossroads @ GI_Forum '08: Proceedings of the Geoinformatics Forum Salzburg*. Edited by Car A, Griesebner G, Strobl J. Heidelberg: Wichmann; 2008:240-245.
10. Schmidt G, Schröder W: Regionalisation of climate variability used for modelling the dispersal of genetically modified oil seed rape in Northern Germany *Ecol Ind* 2011, 11:951-963.
11. Schmidt G, Kleppin L, Schröder W, Breckling B, Reuter H, Eschenbach C, Windhorst W, Höltl K, Wurbs A, Barkmann J, Marggraf R, Thiel M: Systemic risks of genetically modified

organisms in crop production: interdisciplinary perspective.*Gaia* 2009, 18:119-126.

12. Olea RA: *Geostatistics for Engineers and Earth Scientists*. Boston: Kluwer; 1999.
13. Hornsmann I, Pesch R, Schmidt G, Schröder W: Calculation of an Ecological Land Classification of Europe (ELCE) and its application for optimising environmental monitoring networks. In *Geospatial Crossroads @ GI_Forum '08: Proceedings of the Geoinformatics Forum Salzburg*. Edited by Car A, Griesebner G, Strobl J. Heidelberg Wichmann; 2008:140-151.
14. Pesch R, Schmidt G, Schröder W, Aden C, Kleppin L, Holy M: Development implementation and application of the WebGIS MossMet. In *The Geospatial Web: How Geobrowsers, Social Software and the Web 2.0 are Shaping the Network Society*. Edited by Tochtermann K, Scharl A. London: Springer; 2007:191-200.
15. Pesch R, Schmidt G, Schröder W, Weustermann I: Application of CART in ecological landscape mapping: two case studies. *Ecol Ind* 2011, 11:115-122.
16. Schröder W, Schmidt G: Defining ecoregions as framework for the assessment of ecological monitoring networks in Germany by means of GIS and classification and regression trees (CART). *Gate to EHS* 2001, 1-9.
17. Schröder W, Schmidt G, Pesch R: Harmonization of environmental monitoring tools for examination of methodical comparability and spatial representativity. *Gate to EHS* 2003, 1-13.
18. Schröder W, Pesch R, Schmidt G: Analysis of climate change affecting German forests by combination of meteorological and phenological data within a GIS environment. *Sci World* 2007, 7:84-89.
19. Kleppin L, Schmidt G, Schröder W: Cultivation of GMO in Germany: support of monitoring and coexistence issues by WebGIS technology *Environ Sci Eur* 2011, 23:1-11.
20. EIONET [http://terrestrial.eionet.europa.eu/CLC2000]
21. Keil M, Kiefl R, Strunz G: *CORINE Land Cover 2000-Germany. Final Report.*Oberpfaffenhofen: German Aerospace Center, German Remote Sensing Data Center; 2005.

22. EUROSTAT[http://epp.eurostat.ec.europa.eu]
23. Climatic Research Unit (CRU) CL 2.0[http://www.cru.uea.ac.uk/~timm/grid/CRU_CL_2_0.html]
24. New M, Lister D, Hulme M, Makin I: A high-resolution data set of surface climate over global land areas.*Clim Res* 2002, 21:1-25.
25. Englert C, Pesch R, Schmidt G, Schröder W: Analysis of spatially and seasonally varying plant phenology in Germany. In *Geospatial Crossroads @ GI_Forum '08: Proceedings of the Geoinformatics Forum Salzburg*. Edited by Car A, Griesebner G, Strobl J. Heidelberg: Wichmann; 2008:81-89.
26. Oke TR: *Boundary Layer Climates*. London: Routledge; 1987.
27. FAO Digital Soil Map of the World [http://www.fao.org/nr/land/soils/digital-soil-map-of-the-world/en]
28. FAO Harmonized World Soil Databasev 1.1[http://www.fao.org/nr/land/soils/harmonized-world-soil-database/en]
29. Bohn U, Hettwer C, Gollub G: *Anwendung und Auswertung der Karte der natürlichen Vegetation Europas./Application and Analysis of the Map of the Natural Vegetation of Europe*. Bonn: BfN-Skripten (Bundesamt für Naturschutz); 2005.
30. Secretariat of the Convention on Biological Diversity (SCBD): *Cartagena Protocol on Biosafety to the Convention on Biological Diversity*. 2000. Montreal
31. Secretariat of the Convention on Biological Diversity (SCBD): *Global Biosafety: from Concepts to Action. Decisions Adopted by the First Meeting of the Conference of the Parties to the Convention on Biological Diversity Serving as the Meeting of the Parties to the Cartagena Protocol on Biosafety*. 2000. Montreal
32. Vieno M, Toivonen T: *Report of Analysis of Observational Biodiversity Information Needs and Products to Support the European Biodiversity Strategy and Parallel Activities*. Turku: European Network for Biodiversity Information (ENBI); 2005.
33. Gregory RD, Van Strien A, Vorisek P, Meyling AWG, Noble DG, Foppen RPB, Gibbons DW:Developing indicators for European birds. *Philos Trans Roy Soc B Biol Sci* 2005, 360:269-288.
34. Thomas JA: Monitoring change in the abundance and distribution of insects using butterflies and other indicator groups. *Philos Trans R Soc B Biol Sci* 2005, 360:339-357.

35. EuMon database [http://eumon.ckff.si/monitoring/search.php]
36. Schmeller DS, Henry PY, Julliard R, Clobert J, Gruber B, Dziock F, Lengyel S, Nowicki P, Déri E, Budrys E, Kull T, Tali K, Bauch B, Settele J, van Swaay C, Kobler A, Babij V, Papastergiadou E, Henle K: Advantages of volunteer-based biodiversity monitoring in Europe. *Conserv Biol* 23:307-316.
37. Global Biodiversity Information Facility (GBIF) [http://www.gbif.org]
38. European Forest Genetic Resources Programme (EUFORGEN) [http://www.euforgen.org/distribution_maps.html]
39. DIVA-GIS [http://www.diva-gis.org]
40. International Potato Center (CIP) [http://www.cipotato.org]
41. FloraWeb [http://www.floraweb.de]
42. Bundesministerium für Umwelt, Naturschutz und Reaktorsicherheit (BMU): *Umweltbeobachtung. Stand und Entwicklungsmöglichkeiten*. 2000. Bonn
43. Keune H, Mandry P: Harmonization of environmental data. The requirements for developing a consistent view of the environment world-wide. In *Global Monitoring of Terrestrial Ecosystems*. Edited by Schröder W, Fränzle O, Keune H, Mandry P. Berlin, Ernst & Sohn; 1996:51-56.
44. Craig W, Tepfer M, Degrassi G, Ripandelli D: An overview of general features of risk assessments of genetically modified crops. *Euphytica* 2008, 164:853-880.
45. Cao Y, Williams DD, Larsen DP: Comparison of ecological communities: the problem of sample representativeness. *Ecol Monogr* 2002, 72:41-56.
46. Tirler W, Donega M, Voto G, Kahr G: Quick evaluation of long term monitoring samples and its representativeness. *Organohalogen Compounds* 2003, 60:509-512.
47. Bühler C: Biodiversity monitoring in Switzerland: what can we learn for general surveillance of GM crops? *J Verbr Lebensm* 2006, 1:37-41.
48. Leigh RA, Johnston AE: *Long-term Experiments in Agricultural and Ecological Sciences: Proceedings of a Conference to Celebrate the 150th Anniversary of Rothamsted Experimental Station, Held*

at Rothamsted, 14-17 July 1993. Wallingford: CAB International; 1994.

49. Stein A, Ettema C: An overview of spatial sampling procedures and experimental design of spatial studies for ecosystem comparisons. *Agric Ecosyst Environ* 2003, 94:31-47.
50. Wilkinson MJ, Davenport IJ, Charters YM, Jones AE, Allainguillaume J, Butler HT, Mason DC, Raybould AF: A direct regional scale estimate of transgene movement from genetically modified oilseed rape to its wild progenitors. *Mol Ecol* 2000, 9:983-991. PubMed Abstract
51. Hastings DA, Dunbar PK, Elphingstone GM, Bootz M, Murakami H, Maruyama H, Masaharu H, Holland P, Payne J, Bryant NA, Logan TL, Muller J-P, Schreier G, Macdonald JS: *The Global Land One-kilometer Base Elevation (GLOBE) Digital Elevation Model, Version 1.0*. Boulder: National Oceanic and Atmospheric Administration, National Geophysical Data Center; 1999.
52. Ward JH: Hierarchical grouping methods to optimise an objective function. *J Am Stat Assoc* 1963, 58:236-244.
53. Knapp PA, Grissino-Mayer HD, Soule PT: Climatic regionalization and the spatio-temporal occurrence of extreme single-year drought events (1500-1998) in the Interior Pacific Northwest, USA. *Quaternary Res* 2002, 58:226-233.
54. Shwartz M, Payne SMC, Restuccia JD, Ash AS: Does it matter how small geographic areas are constructed? Ward's algorithm versus the plurality rule. *Health Serv Outcomes Res Methodol* 2001, 2:5-18.
55. Zhou Y, Narumalani S, Waltman WJ, Waltman SW, Palecki MA: A GIS based spatial pattern analysis model for ecoregion mapping and characterization. *Int J Geogr Inform Sci* 2003, 17:445-462.
56. Glemnitz M, Wurbs A, Roth R: Derivation of regional crop sequences as an indicator for potential GMP dispersal on large spatial scales. *Ecol Indicat* 2011, 11:964-973.
57. Middelhoff U, Reuter H, Breckling B: GeneTraMP, a spatio-temporal model of the dispersal and persistence of transgenes in feral, volunteer and crop plants of oilseed rape and related species. *Ecol Indicat* 2011, 11:974-988.

58. Reuter H, Schmidt G, Schröder W, Middelhoff U, Pehlke H, Breckling B: Regional distribution of genetically modified organisms (GMOs)--up-scaling the dispersal and persistence potential of herbicide resistant oilseed rape (*Brassisca napus*). *Ecol Indicat* 2011, 11:989-999.

Chapter 8

Cultivation of Gmo in Germany: Support of Monitoring and Coexistence Issues by Webgis Technology

Lukas Kleppin, Gunther Schmidt, and Winfried Schröder

University of Vechta, 49364 Vechta, Germany

ABSTRACT

Background

In Germany, apart from the Amflora potato licensed for cultivation since March 2010, Bt-maize MON810 is the only genetically modified organisms (GMO) licensed for commercial cultivation (about 3,000 ha in 2008). Concerns have been raised about potential adverse environmental impacts of the GMO and about potential implications on the coexistence between conventional and genetically modified production. These issues should be considered on a regional base. The objective of this article is to describe how GMO monitoring that is required after risk assessment and GMO release can be complemented by a Web-based geoinformation system (WebGIS). Secondly, it is also described how WebGIS techniques might support coexistence issues with regard to Bt-maize cultivation and conservation areas. Accordingly, on the one hand, the WebGIS should enable access to relevant geodata describing the receiving environment, including information on cultivation patterns and conservation areas containing protected species and habitats. On the other hand, metadata on already established environmental monitoring networks should be provided as well as measurement data of the intended GMO monitoring. Based on this information and based on the functionality provided by the WebGIS, the application helps in detecting possible environmental GMO impacts and in avoiding or identifying coexistence problems.

Results

The WebGIS applies Web mapping techniques to generate maps via internet requests and offers additional functionality for analysis, processing and publication of selected geodata. It is based on open source software solely. The developments rely on a combination of the University of Minnesota (UMN) MapServer with the Apache HTTP server, the open source database management systems MySQL and PostgreSQL and the graphical user interface provided by Mapbender. Important information on the number and the location of Bt-maize fields were derived from the GMO location register of BVL. The "WebGIS GMO Monitoring" provides different tools allowing for the application of basic GIS techniques as, for instance, automatic or interactive zooming, distance measurements or querying attribute information from selected GIS layers. More sophisticated GIS tools were implemented additionally, e.g. a buffer function which enables generating buffers around selected geo-objects like Bt-maize fields. Finally, a function for intersection of different maps was developed. The WebGIS comprises information on the location of all Bt-maize fields in Germany according to the official GMO location register of the Federal Office of Consumer Protection and Food Safety between 2005 and 2008. It facilitates, amongst others, access to geodata of GMO fields and their surroundings and can relate them with additional environmental data on climate, soil, and agricultural patterns. Furthermore, spatial data on the location of flora-fauna-habitats and environmental monitoring sites in the federal state of Brandenburg were integrated.

The WebGIS GMO monitoring was implemented according to the concept for an "Information System for Monitoring GMO" (ISMO) which was designed on behalf of the German Federal Agency for Nature Conservation. ISMO includes hypotheses-based ecological effects of GMO cultivation and suggests checkpoints for GMO monitoring to test whether impacts may be observed in the receiving environment.

In contrast to the public GMO register, the WebGIS GMO monitoring enables mapping of GMO fields and provides relevant geodata describing environmental and agricultural conditions in their neighbourhood of the cultivation sites as well as information derived from monitoring sites. On this basis, spatial analyses should be enabled and supported, respectively. Further, the WebGIS GMO monitoring supplements PortalU which, in Germany, is the technical realisation of the Infrastructure for Spatial Information in Europe directive (Directive 2007/2/EC) released by the EU in 2007.

Conclusions

The article should have shown how to support and complement GMO monitoring with the help of the WebGIS application. It facilitates co-operation and data access across spatial scales for different users since it is based on internet technologies. The WebGIS improves storage, analysis, management and presentation of spatial data. Apart from the improved flow of information, it supports future long-term GMO monitoring and modelling of the dispersal of transgenic pollen, for instance. Additional information (e.g. data on wind conditions or soil observation sites) provided by the WebGIS will be helpful to determine representative monitoring sites for detecting potential GMO impacts by means of monitoring or modelling. Thus, the WebGIS can also serve as part of an early warning system. In the near future, the integration of locations of all Bt-maize fields in Germany into the WebGIS as a continuous task should be automatised. Additionally, a methodology should be developed to detect maize fields by means of remote sensing data to manage coexistence problems on the basis of actual field patterns.

BACKGROUND

Genetic engineering was introduced to improve plant breeding. It enables to establish new varieties of plant species with specific input and output traits [1]. The cultivation of GMO aims at increasing yield, but also to improve product quality [2,3]. Input traits include resistances against different herbicides or insect pests and viruses. Output traits aim to improve the quality of agricultural products, e.g. increasing fibre or lowering the fat content. Worldwide, the cultivation of GMO increased from 1.7 Mio ha in 1996 up to 134 Mio ha in 2009 [4]. According to the agricultural statistical survey 2009, for example in the USA, 90% of the cropland is used to cultivate GMO varieties of soy or cotton. In the USA, the percentage of genetically modified (GM) maize is already 85%. In Germany, GMO (99% Bt-maize MON810) were cultivated from 2005 until 2008 with a total number of 239 fields and a total acreage of 3,171 ha in 2008.

In contrast to the contained use of GM products in medicine, the introduction of GMO in agricultural ecosystems may cause unwanted, uncontrollable and irreversible impacts.

According to EU Directive 2001/18/EC, plant breeders willing to introduce GMO on the market have to accomplish a notification process including an environmental risk assessment (ERA) and a monitoring plan to the competent national and European authorities [5]. This regulatory framework is intended to implement the precautionary principle and to enable handling potential adverse environmental effects still remaining after the ERA [6]. The aim of the EU Directive 2001/18/EC is to safeguard human health and the environment and to restrict the use of GMO so that no unacceptable risks or hazards can emerge [7]. The risk assessment is based on empirical studies with small spatial extent, encompassing laboratories tests, greenhouse experiments, small-scale field release or commercial-scale field release [8-10]. Though, there remains a wide range of uncertainty with small plot and laboratory studies. According to scientific hypotheses, adverse effects are examined in the ERA. However, ERA concentrates at the small-scale level, thus, large-scale effects are difficult to assess. Thus, monitoring of GMO at the landscape scale is required after the GMO have been released to detect adverse environmental effects at regional or larger scale [11]. Accordingly, the EU Directive 2001/18/EC [5] on the deliberate release of GMO into the environment stipulates assessment of direct and indirect effects of GMO on humans and the environment by case-specific monitoring and general surveillance. The latter has to be performed to detect potential unanticipated adverse effects whereas case-specific monitoring is set up to reduce substantial uncertainties in relevant risk scenarios identified in the ERA [5]. In Germany, the Federal Nature Conservation Agency suggests how to implement a monitoring of GMO. Three core issues have to be covered: (1) documentation of exposure, (2) monitoring impacts of the specific GMO and (3) large-scale and long term-effect relationships [12]. The results of GMO monitoring contribute to decisions regarding, e.g. further approval or refusal of the GMO or additional precautions during cultivation. In this context, GMO monitoring provides the basis for an early warning system to react at an early stage in case of reported adverse effects and decide upon counter measures. Relevant topics have to be considered for both, case-specific monitoring and general surveillance, which are, for example, (a) combinatory effects of several genetic modifications accumulating in individual plants of a crop species such as multiple resistances in oilseed rape [13,14], (b) effects of different Bt-toxins on susceptible butterfly populations [15-17] or (c) long-term effects due to changes in farming practices [18]. The necessity of monitoring adverse GMO effects can be pointed out by means of a few indications, for example, enhanced mortality of non-target organisms [6], hybridisation with related species [19] or neighbouring non-GM crops [20] and adverse agricultural practice changes [21]. According to EU Directive 2001/18/EC [5], a set of appropriate monitoring parameters has to be defined which are described in the guidelines for GMO monitoring as, for instance, published by the Association of German Engineers [22]. These obligate test items have to be considered when integrating and compiling data from already existing environmental monitoring networks [23-25].

In this context, a Web-based geographical information system (WebGIS) is appropriate to build up a data infrastructure for GMO monitoring and data exchange [26]. The objective of the article at hand is to describe how to complement and support GMO monitoring by the implementation of a WebGIS as suggested by Aden et al. [25]. The WebGIS enables access to relevant geodata like basic environmental information, existing monitoring networks related to GMO issues, details on GMO fields and information on protected areas as well as tools for collecting, processing and mapping monitoring results. Implemented GIS tools that do not require any additional software but an Internet browser at the client's computer should help in assessment of possible GMO impacts in a spatially discriminated context. On that score, the WebGIS can facilitate the approval process. Secondly, it could be used to manage coexistence of GMO, conventional and organic farming as well as with nature conservation issues by detecting or avoiding possible conflicts already during planning stage [27]. Moreover, the Web-based application will provide spatial information on the locations of the Bt-maize fields which can be used for modelling cross-pollination of GM maize pollen at field scale, for instance, to check, e.g. whether distance regulations between Bt-maize and conventional maize fields are sufficient or not [28].

MATERIALS AND METHODS

Open Source Software and Standards

The use of proprietary software is being determined by licences and copyrights; annual license fees may be imposed. Sharing or modification of this software is strictly forbidden. Due to the business concept of proprietary software, the source code is not accessible [29]. Open source software offers an approved alternative to proprietary software. However, there is no guarantee that the open source software is working properly. Compared to proprietary software, open source products are prescribed to be free of charge and the source code is disclosed and free for modifications. Open source software is not confined to private use, but is adopted from business companies, public facilities as well as from authorities. It is used in all fields of information technology, for instance, as operating system (Linux, HostGIS), complementary software (hypertext transfer protocol (HTTP) Server, CMS, MapServer, WebGIS-Clientsuites), and independent GIS software (GRASS-GIS, JUMP) [29]. Open source is specified by several criteria of the Open Source Initiative [30,31].

Open source software used to build up the Web application described in the article at hand follows the standards of the Open Geospatial Consortium (OGC). This is an international organisation composed of business companies, universities and authorities. The OGC releases standards for interfaces to process various types of geodata via Internet. Standards and specifications are supposed to ensure interoperability between map services located anywhere in the world and to provide access to complex spatial information. The EU directive Infrastructure for Spatial Information in Europe (INSPIRE) [32] and the German PortalU [33] already comply with these standards.

System Architecture

Based on open source software and in accordance with the INSPIRE standards, we developed the "WebGIS GMO Monitoring". To this end, a server programme was used which provides the functionality of a "spatial" communication. Our recent developments rely on a combination of the UMN MapServer with the Apache HTTP server. The main function of the HTTP server relies on the communication with Web clients. Map servers are components that perform queries and analyses of both raster and vector data and generate and display maps in a uniform projection defined by the user. We then installed the database management system PostgreSQL enhanced with the spatial extension PostGIS. Open source database systems like MySQL and PostgreSQL are capable to save and process spatial information and related attributes in additional libraries (MyGIS, PostGIS). The spatial extension PostGIS acts as GIS back end which allows performing basic GIS operations on geodata without expensive programming. The integration of the GIS back end GRASS is an essential part of the current work. The WebGIS interactively enables advanced GIS techniques and geodata analyses. For this purpose, the user only needs a Web browser (Mozilla, MS Internet Explorer) but no additional GIS software. Finally, we installed the WebGIS-Client Suite Mapbender by CCGIShttp://www.mapbender.org which provides the user interface. The open source product offers various tools for navigation within maps, retrieval of metadata and queries of map contents[34]. Moreover, it is possible to integrate remote Web Map Services to build up a more extensive data infrastructure for environmental monitoring issues.

GMO Location Register

The Federal Office of Consumer Protection and Food Safety (*Bundesamt für Verbraucherschutz und Lebensmittelsicherheit*, BVL) is the competent authority charged with the enforcement of the Genetic Engineering Act (*Gentechnikgesetz*, GenTG) and the legislation of the European Union. The BVL, correspondingly, assesses notifications for the experimental use of GMO and also gives advice to the Federal Government as well as to the Federal States and their bodies on issues of biological safety in genetic engineering. The BVL maintains the GMO location register [35]http://www.bvl.bund.de as well as the GMO notification register, serving as an information platform on GMO release for the public. The BVL is committed to record information on GMO cultivation in the register by the EU Directive 2004/204/EC [36]. This is to improve monitoring of possible negative long-term effects with regard to environment, human and animal health. Additionally, the GMO location register should assure transparency and should help adjacent farmers to cultivate GM crops and non-GM crops without cross-pollination (coexistence). The GMO location register contains the identification numbers (ID) of GMO fields related to the *Amtliches Liegenschaftskataster* (ALK). However, the GMO location register is not linked with the ALK and has only very limited options for cartographic visualisation, i.e. it is only possible to map the cultivation of Bt-maize at the level of municipalities in terms of density maps [37]. A visualisation of Bt-maize fields is not possible by the location register and, thus, it is not possible to identify single GMO fields by spatial queries or mapping.

The application WebGIS GMO monitoring improves these techniques and provides corresponding information to implement required monitoring issues (GenTG, chapter 3, 15). As a first step of development and implementation, the WebGIS was designed only for the Federal State of Brandenburg. The localisation of the GMO fields in Brandenburg was enabled by identification of the land parcels where the Bt-maize was cultivated using the ID field of the ALK listed in the GMO location register. Difficulties arise when no public cadastre (ALK) is available for free to spatially reference according GM maize fields (see "Conclusions").

RESULTS

Database

Geodata having been integrated in the WebGIS application are essential for GMO monitoring issues because in various ways they can help detect possible impacts or coexistence problems due to GMO cultivation. The application provides maps on land use patterns of CORINE Land Cover [38], on ecological landscape units [39] and on ecoregions [40] as well as satellite images of Northern Germany, phenological data on maize plants and averaged measurements on precipitation (1961-1990), temperature (1961-1990), sunshine duration (1961-1990), wind direction and evaporation rate compiled from the German Weather Service (DWD). Furthermore, maps on cultivation intensity of several crops at district level derived from agricultural statistics (*Statistik lokal* 1999, 2003, 2007)[41] and data on Bt-maize cultivation derived from the public GMO register were integrated. In addition to the developmental stage of the WebGIS as published by Kleppin et al. (2008) [42], supplementary data were integrated in the WebGIS GMO monitoring: information on the location of fauna-flora-habitats (FFH) in the federal state of Brandenburg including a list on protected species[43], data on monitoring programmes in Brandenburg with regard to long-term soil observation sites, groundwater and surface water observation sites as well as monitoring sites within biosphere reserves[44] including a list of analysed parameters. Finally, the database was updated with information on the occurrence of the European corn borer (*Ostrinia nubilalis*) from 2005 until 2007 being the target organism for the introduction of Bt-maize. All geodata and according attributes are described by metadata which can be modified or completed if necessary. The WebGIS administrator is authorised to decide whether actual geodata may be downloaded by user request. By this, users get distinct access rights for predetermined information.

The WebGIS GMO Monitoring

The WebGIS GMO monitoring provides a graphical user interface based on the Mapbender software (Figure 1). A tool bar allows applying basic GIS techniques (see Figure 1, item 3), for instance, automatic or interactive zooming, distance measurements or querying attribute information from selected GIS layers. A detailed map including a scale bar and navigation buttons show the selected layers (see Figure 1, item 5). A small-scale reference map depicts the geographical location of the selected area displayed in the detailed map (see Figure 1, item 2). The layer tool enables management of geo-objects (see Figure 1, item 1). By activating the checkboxes, each layer is drawn in the map window (left checkbox in item 1) or attribute queries can be enabled (right checkbox in item 1). Corresponding to the chosen layers, legends are generated automatically (see Figure 1, item 4). The selected layers 'Cultivation 2008' (A) and 'GMO sites' (B) displayed in Figure 1 show (A) the cultivation area of Bt-maize fields for each municipality in 2008, and (B) in detail single Bt-maize fields in Brandenburg which were registered by the BVL in 2008. The map on Bt-maize fields can be complemented by displaying additional geodata, like, for instance, maps on land use patterns or ecoregions as, e.g., published by Schröder and Schmidt (2001) [40]. Additionally, maps on the location of nature reserves can be overlaid with locations of Bt-maize fields. By clicking on the layer's name in the WebGIS application, available metadata describing source, date of origin and other relevant information on the data set are listed in tables.

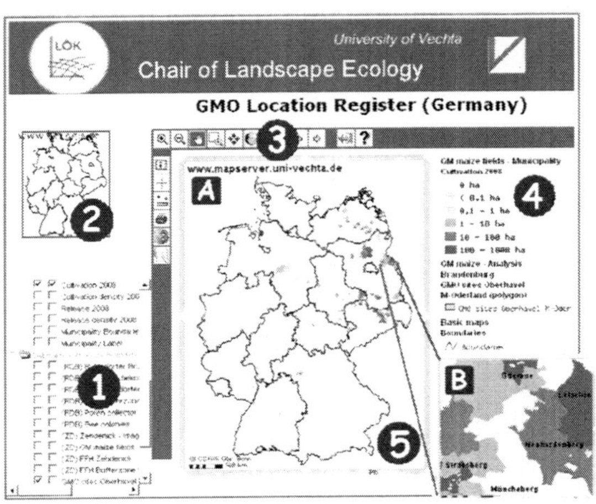

Figure 1: WebGIS GMO monitoring displaying percentage of Bt-maize fields. (a) In relation to total maize cropland and (b) a detailed map on the allocation of Bt-maize fields in Brandenburg (yellow).

Beyond the developmental stage of the WebGIS as reported by Kleppin et al. (2008) [42], the WebGIS GMO monitoring was improved by the implementation of sophisticated GIS tools. A buffer function allows generating buffers around selected geo-objects like, for instance, Bt-maize fields (Figure 2B). Another function ("contain") allows listing of all geo-objects being located within a certain buffer zone (Figure 2D). An "intersect" function (Figure 2C) can be used for spatially relating different layers. Two special intersect cases were realised, such as "clip" and "union". "Clip" can be used to cut out features of one layer with one or more features of another layer. The function "union" calculates the geometric intersection of all features of two layers. The output features will then have the attributes of both layers. Further, it is possible to calculate distances between geo-objects (Figure2E) and, finally, a query tool was implemented to identify distinct GMO fields. It is also possible to generate buffer zones around single or several (Bt-)maize fields in a given municipality by specifying a buffer name and the desired extent of the buffer zone. The username is necessary to generate unique names for both the new layer (see Figure 2B). While the buffer zone is calculated, the map file, which defines the layout of the new geo-object, is generated automatically too, and integrated into the user interface of the Mapbender software ("The WebGIS GMO monitoring"). Additionally, the new buffer zone as well as the respective SRID (Spatial Reference Identifier) and the type of the geometry are registered dynamically in the geodatabase. For displaying the new layer it is necessary to update the webpage (Figure 3). The new buffer-layer can be intersected with other geodata stored in the geodatabase (Figure 2F). Further, an according template file provides specific information which describes the selected area or location by coordinates, name, size, etc. After log out, all files and geodata generated before are deleted in order to save storage capacity. Additional extensions for printing maps or downloading the individually generated files are under construction.

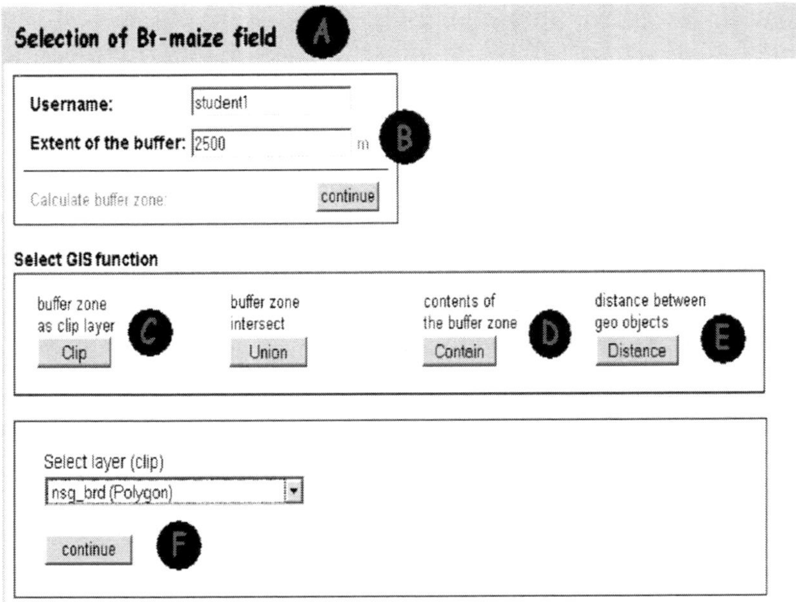

Figure 2: GIS operations for analysing geo-objects (Bt-maize fields).

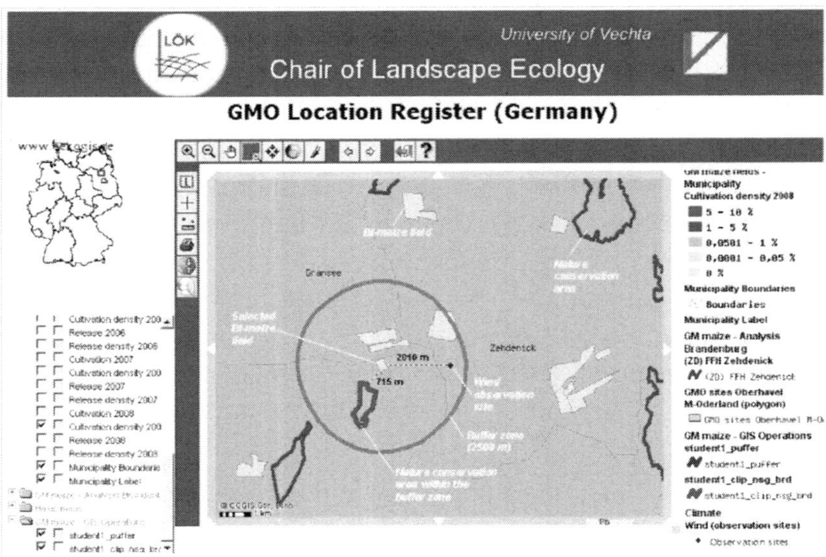

Figure 3: WebGIS GMO monitoring showing the visualisation of different geodata in the layer folder (GM maize-GIS Operations).

As an example, in Figure 2A, a certain Bt-maize field is selected to generate a buffer zone of 2,500 m around this field. As a result, the extent of the buffer appears in the map as a blue polygon (see Figure 3). In the next step, the user extracts geo-objects from the FFH layer by clipping with the buffer layer generated before (see Figure 2C, F). In the result, one single FFH area is highlighted (red outline) being located within the buffer zone (see Figure 3). Additionally, the extracted FFH area is linked to a query template to provide specific information, for instance, on protected species housed in this FFH area. This spatial investigation whether the Bt-maize fields are within or near a conservation area is relevant since protected non-target organisms might be affected by toxins produced by Bt-maize or a change in biodiversity might be induced. Furthermore, it is possible to calculate the distance between the selected Bt-maize field and the respective conservation area (see Figure 2E) and to identify other relevant geodata located within the buffer zone (see Figure 2D).

In case local authorities plan to conduct a case-specific GMO monitoring, buffer zones around all Bt-maize fields of the respective municipality might be generated at first. In a second step, it could be checked automatically whether monitoring sites of related environmental monitoring networks ("Database") are located within the buffer zones. Regarding the respective GMO, it could be checked in detail what measurements are taken at these sites in order to support analysis of possible adverse effects. For instance, data on wind conditions can be evaluated in order to determine favourable sites for technical pollen samplers [45]. Projected GM pollen loads help in assessing risks for non-target organisms (NTO) occurring in the vicinity of GMO fields. In this context, Rosi-Marshall et al. (2007)[46] found out in laboratory feeding trials that consumption of Bt-corn byproducts reduced growth and increased mortality of NTO stream insects.

Another benefit of the WebGIS GMO monitoring refers to coexistence issues. Generally, coexistence refers to the choice of consumers and farmers between conventional, organic and GM crop production. Thus, the aim is to accomplish a spatial segregation between GM and non-GM production at the landscape level which helps to avoid cross-pollination and seed contamination. Similarly, conflicts between GMO cultivation and protection goals concerning conservation reserves have to be avoided. By use of the WebGIS, farmers cultivating conventional maize are enabled to check distances to adjacent Bt-maize fields with regard to distance regulations defined in the amendment of the GenTG (150 m to conventional fields, 300 m to organic fields). This also applies to protected areas with respect to nature conservation issues (800 m in the federal state Brandenburg).

The localisation of regions where Bt-maize can be cultivated without impairing conventional maize fields or nature reserves is a challenging task. In a GIS-based approach, conventional maize fields and conservation areas have to be buffered in accordance to existing distance regulations. The cropland outside the buffered area would be eligible for Bt-maize cultivation [47]. Furthermore, all FFH conservation areas in Brandenburg are documented by subjects of protection (e.g. endangered species). In order to identify all the FFH conservation areas which might probably be affected by pollen dispersal, it is necessary to generate a buffer of 800 m, as defined by the federal authorities, around these areas. In the next step, the according buffer zones must be intersected with the geometries of the Bt-maize fields to identify whether some of these Bt-maize fields are located within the respective buffer zone. Afterwards, it can be tested whether any endangered species (NTOs) occur in the respective protection areas which might be exposed to Bt-maize pollen. Laboratory tests have shown that Bt toxins may influence NTOs in growth and physical condition [48,46].

DISCUSSION

GMO monitoring should take place in areas exposed to GMO, preferably cultivated fields and their environment, but should include also regions with no or unknown GMO exposure. On a case-by-case basis depending on the GMO characteristics, the selected indicators, checkpoints and related analytical methods should consider relevant different spatial and temporal scales [49,22]. Hence, the monitoring of ecological effects of GMO must be standardised with regard to parameters, methods, survey intervals and sites so that data are comparable in terms of measurement methods and, thus, can be analysed statistically and interpreted meaningfully [22]. This comprises standards concerning molecular-biological detection methods, vegetation mapping and faunistic surveys to evaluate changes in population density and behaviour of endangered species, for example. This standardisation is to ensure a Germany-wide comparability of sampling data and to provide legal certainty for the user [50]. Accordingly, the WebGIS GMO monitoring should support realisation of particular parts of the guideline VDI 4330 [22]: "Monitoring the ecological effects of genetically modified organisms - Basic principles and strategies" (VDI 4330, part 1), "Pollen monitoring: Pollen sampling using pollen mass filters (PMF) and Sigma-2 samplers" (VDI 4330, part 3), "Pollen monitoring: Biological sampling by honey bees" (VDI 4330, part 4). In this context, Reuter et al. (2006, 2010) [23,24] developed a concept of an information system for GMO monitoring (ISMO). The database concept encompasses three components: The "Knowledge Database" comprises information related to different levels of biological organisation being affected by GMO cultivation. Therein, scientific hypotheses regarding ecological effects of GMO as well as checkpoints for monitoring possible impacts were described in detail. The "Monitoring Database" should provide GMO monitoring data and interfaces to existing environmental information systems being of relevance for GMO monitoring issues. The WebGIS GMO monitoring is designated to be part of the monitoring database enabling data retrieval, mapping and analysis of relevant monitoring data and geodata. The "Administrative Database" structures all data necessary for the approval process. ISMO enables support by competent authorities in the notification process and post market monitoring of environmental effects [24]. Checkpoints defined by ISMO were used to compile and integrate appropriate environmental monitoring programmes in the WebGIS GMO monitoring.

Compared with the public register of the BVL, advantages of the WebGIS GMO monitoring are obviously the possibility to map registered GMO fields as well as to perform spatial analyses by additional relevant geodata useful for GMO monitoring issues and environmental risk assessment. The use of licence-free open source software for assembling the application is another advantage compared with the public register of the BVL which is based on proprietary software. The WebGIS GMO monitoring is not intended to compete with the public register of the BVL, but it serves as a supplement for more transparency regarding the localisation and management of single GMO fields and agricultural patterns.

The Federal Nature Conservation Agency provides another WebGIS application [51] which allows displaying Natura 2000 reserves as well as predefined buffer zones of 1,000 m around them. An interactive query offers additional information on the respective nature reserve, like name and site code. Specific information on protected species in general or species that might be affected by GMO cultivation individually is not provided.

Another application called "Risk Register Genetic Engineering Agriculture" [52] displays, for instance, all Bt-maize fields cultivated in 2009 and 2010 in Germany. The respective field geometries were derived by using Google Maps. Additional thematic maps were integrated on the basis of the official GMO location register of the BVL displaying static density maps of GMO cultivation on different administrative levels and for different periods and crops. However, this application just visualises GMO fields, whereas the WebGIS GMO monitoring additionally enables performing GIS procedures. Furthermore, interactive dynamic generation of buffers and intersection with additional geodata enhance the WebGIS functionalities in terms of spatial analysis. For instance, it is possible to intersect data on Bt-maize fields with additional geodata like related monitoring sites or distribution maps of the corn borer as being the target object for Bt-maize cultivation. Furthermore, the WebGIS GMO monitoring facilitates linkage to PortalU [53] as being the German realisation of the European INSPIRE directive [33] which aims at "establishing an infrastructure for spatial information in Europe to support Community environmental policies, and policies or activities which may have an impact on the environment". Accordingly, data from the WebGIS GMO monitoring will enhance the database of PortalU and enable remote geodata access without implementation of a local GIS software at the client PC.

Compared to the work published by Kleppin et al. (2008) [42], the database was complemented by additional geodata, e.g., on environmental monitoring networks and the respective information on measurement parameters. Apart from that, the WebGIS GMO monitoring was optimised and improved by the implementation of additional sophisticated GIS techniques including buffer and intersect tools. However, long-term risks of GMO cultivation are difficult to assess, in particular, because possible impacts depend on spatially varying conditions [54]. Anticipating risks is often hampered by limitations in scientific knowledge or in availability of data, in particular, in cases where a complex process of change is continuing (e.g. climate change) or a new technological context is added to an established interaction network. An increasing amount of information can be accessed via the Internet. Particular in recent years, attention has focused on the presented WebGIS technology which enables compilation and access to data, e.g., affecting the dispersal of GMO, such as wind speed and direction. However, GIS is not only used for pre-event vulnerability assessment but can be used also for improving preparedness, mitigation, monitoring and response plan activities. Thus, the use of WebGIS provides instructive links with administrative, socioeconomic and other data, and enhances communication of the results to policy makers and the public. This communication dimension is fundamental - local people need to incorporate risk awareness into their culture [55].

CONCLUSIONS

According to Wilkinson et al. (2000) [56] and Züghart and Breckling (2003) [57] criteria for selecting monitoring sites and regions include 1) representativeness of sites cultivated with specific GMO, 2) representativeness of ecological regions containing the spectrum of relevant indicators, 3) availability of sites already monitored within other environmental programmes, and 4) areas with environmental conditions facilitating spread or survival of GMO. The WebGIS GMO monitoring supports this task by providing data on the distribution of GMO fields as well as on the distribution of monitoring sites of different environmental monitoring programs and, thus, helps in selecting appropriate monitoring sites. Furthermore, the article at hand demonstrates that the use of the WebGIS GMO monitoring is a useful and efficient tool to assess the individual and spatial risk potential before and during GMO release since it can be used to identify coexistence problems between Bt-maize and conventional maize cultivation on the one hand and between Bt-maize cultivation and conservation issues on the other hand.

Since in the future number and location of GMO fields might change considerably, the integration of geometries of Bt-maize fields into the WebGIS should be improved by an automation of the update procedure. Difficulties arise when no free Web services on ALK data are available to locate the respective GMO fields precisely. This is the case for about one third of all federal states in Germany. A possible solution is to compile the information directly from land registry offices. But this causes additional costs and is very time consuming. In this context, access to the 'Integrated Administration and Control System' (IASC, in German: InVeKoS = Integriertes Verwaltungs- und Kontrollsystem) would be a better and more efficient solution. Referring to this, in 1992 in the course of the reform of the 'Common Agriculture Policy' (GAP) the implementation of an 'IASC' was decided. It was introduced in Germany as per regulation no. 1782/2003 on December 3th 2004 (BGBl. 1 p. 3194) in order to define cultivation premiums. However, a nationwide information system on spatial and temporal cultivation patterns integrating agricultural data of all federal states has not been established, yet. In completion with information provided by the official GMO location register, such information platform could be used for detection of potential conflict regions with regard to conventional and Bt-maize cultivation. Since access to these data is not possible so far, future work aims at detection of maize fields by remote sensing data to integrate these data into the WebGIS application.

Furthermore, pollen dispersal plays an important part in the spread of GMO. Thus, currently the dispersion model AUSTAL2000 developed by the German Environmental Protection Agency [58] is being implemented in the WebGIS GMO monitoring. The Lagrange particle model considers time-dependent emissions from road and industrial sources. Modification of this software should enable simulation of Bt-maize pollen dispersal to quantify pollen load into conservation areas or conventional maize fields. Further, the dispersion model could help to establish a pollen monitoring network based on technical samplers or biological sampling by bees with respect to VDI 4330, parts 3 and 4 described by Hofmann et al. (2010) [45].

AUTHORS' CONTRIBUTIONS

LK developed the WebGIS GMO monitoring and drafted the manuscript. GS composed the section Backgrounds and participated in the development of useful GIS operations. WS concentrated on the chapters Discussion and Conclusions. All authors read and approved the final manuscript.

REFERENCES

1. Pickardt T, de Kathen A: *Gentechnisch veränderte Pflanzen mit neuen oder verbesserten Qualitäts- und Nutzungseigenschaften: Futtermittel und rohstoffliefernde Nutzpflanzen, Pflanzen zur Bodensanierung und Zierpflanzen*. BioTechConsult. Berlin; 2004:1-107.
2. Squire GR, Hawes C, Begg GH, Young MW: Cumulative impact of GM herbicide tolerant cropping on arable plants assessed through species-based and functional taxonomies. *Environ Sci Pollut Res* 2008, 16:85-94.
3. Wünn J: Landwirtschaft und Ernährung: Was bringen die neuen Entwicklungen für Bauern und Verbraucher? In *Pflanzenbiotechnologie in der Schweiz. Ein Jahr nach der "Gentechfrei-Initiative". Bericht zur Fachtagung des Zürich-Basel Plant Science Center* Edited by Kohler S, Maranta A, Sautter Ch. Zürich. 2007, 6-9.
4. James C: Global Status of Commercialized Biotech/GM Crops: 2009 - The first fourteen years, 1996 to 2009. [http://www.isaaa.org] *webcite ISAAA Executive Summary, Brief No. 41. Ithaca, NY*
5. EC: EU Directive 2001/18/EC of the European Parliament and of the council 12 March 2001 on the deliberate release into the environment of genetically modified organisms and repealing Council Directive 90/220/EE. 2001, L106:1-38.
6. Sears MK, Hellmich RL, Stanley-Horn DE, Oberhauser KS, Pleasants JM, Mattila HR, Siegfried BD, Dively GP: Impact of Bt corn pollen on monarch butterfly populations: a risk assessment. *PNAS* 2001, 98:11937-11942.
7. EFSA-Q-2003-005: Guidance document of the scientific panel on genetically modified organisms for the risk assessment of genetically modified plants and derived food and feed (Question No EFSA-Q-2003-005). Adopted on 24 September 2004, updated on 7 December 2005, final edited version of 28 April 2006, May 2006
8. Chapman MA, Burke JM: Letting the gene out of the bottle: The population genetics of genetically midified crops.*New Phytologist* 2006, 170:429-443.

9. Devaux C, Lavigne C, Austerlitz F, Klein EK: Modelling and estimating pollen movement in oilseed rape (*Brassica napus*) at the landscape scale using genetic markers. *Mol Ecol* 2007, 16:487-499.

10. Spök A, Hofer H, Lehner P, Valenta R, Stirn S, Gaugitsch H: Risk assessment of GMO products in the European Union. Toxicity assessment, allergenicity assessment and substantial equivalence in practice and proposals for improvement and standardisation. *Wien* 2005. [*Berichte*, BE-253]

11. Craig W, Tepfer M, Degrasi G, Ripandelli : An overview of general features of risk assessments of genetically modified crops. *Euphytica* 2008, 164:853-880.

12. Züghart W: Long-term and large-scale effects of genetically modified organisms require specific environmental monitoring designs. In *Implications of GM-Crop Cultivation at Large Spatial Scales*. Edited by Breckling B, Reuter H, Verhoeven R. Frankfurt: Peter Lang; 2008:81-85. [Theorie in der Ökologie 14].

13. Beckie HJ, Warwick SI, Nair H, Seguin-Swartz G: Gene flow in commercial fields of herbicide-resistant canola (*Brassica napus*). *Ecol Appl* 2003, 13:1276-1294.

14. Knispel AL, Mclachlan SM, Van Acker RC, Friesen LF: Gene flow and multiple herbicide resistance in escaped canola populations. *Weed Sci* 2008, 56:72-80.

15. Dively GP, Rose R, Sears MK, Hellmich RL, Stanley-Horn DE, Calvin DD, Russo JM, Anderson PL: Effects on monarch butterfly larvae (Lepidoptera: Danaidae) after continuous exposure to Cry1ab-expressing corn during anthesis. *Environ Entomol* 2004, 33:116-1125.

16. Lang A, Vojtech E: The effects of pollen consumption of transgenic Bt maize on the common swallowtail, Papilio machaon L. (Lepidoptera, Papilionidae). *Basic Appl Ecol* 2006, 7:296-306.

17. Losey JE, Rayor LS, Carter ME: Transgenic pollen harm monarch larvae. *Nature* 1999, 399:214.

18. Graef F: Agro-environmental effects due to altered cultivation practices with genetically modified herbicide-tolerant oilseed rape and implications for monitoring. A review. *Agronom Sustain Dev* 2009, 29:31-42.

19. Lefol E, Fleury A, Darmency H: Gene dispersal from transgenic crops. II. Hybridization between oilseed rape and the wild hoary mustard. *Sex Plant Reprod* 1996, 9:189-196.

20. Rieger MA, Lamond M, Preston C, Powles SB, Roush RT: Pollen-mediated movement of herbicide resistance between commercial canola fields. *Science* 2001, 296:2386-2388.

21. Graef F, Stachow U, Werner A, Schütte G: Agricultural practice changes with cultivating genetically modified herbicide-tolerant oilseed rape. *Agricult Syst* 2007, 94:111-118.

22. VDI (Verein Deutscher Ingenieure): *Beobachtung ökologischer Wirkungen gentechnisch veränderten Organismen - Gentechnisch veränderte Pflanzen - Grundlagen und Strategie*. VDI 4330, Blatt 1. Düsseldorf; 2006.

23. Reuter H, Verhoeven R, Middelhoff U, Breckling B: Information system for the monitoring of genetically modified organisms (GMO) - ISMO. *J Verbr Lebensm* 2006, 1:89-91.

24. Reuter H, Middelhoff U, Graef F, Verhoeven R, Batz T, Weis M, Schmidt G, Schröder W, Breckling B: Information system for monitoring environmental impacts of genetically modified organisms. [http://dx.doi.org/10.1007/s11356-010-0334-y] *webcite Environ Sci Pollut Res* 2010.

25. Züghart W, Benzler A, Berhorn F, Sukopp U, Graef F: Determining indicators, methods and sites for monitoring potential adverse effects of genetically modified plants to the environment: the legal and conceptional framework for implementation. *Euphytica* 2008, 164:845-852.

26. Aden C, Schmidt G, Schröder W: Ein webbasiertes Geografisches Informationssystem für das Monitoring gentechnisch veränderter Organismen. In *GVO-Monitoring vor der Umsetzung. Bonn* Edited by Breckling B, Dolek M, Lang A, Reuter H, Verhoeven R. 2007, 97-112. [Bundesamt für Naturschutz (Series Editor): *Naturschutz und Biologische Vielfalt*, 49]

27. Schmidt G, Schröder W: Auswahl repräsentativer Standorte zur Modellierung der Ausbreitung von gentechnisch veränderten Pflanzen in Nord-Deutschland. *Umweltwiss Schadst Forsch* 2008, 20:9-23.

28. Breckling B, Reuter H, Middelhoff U, Glemnitz M, Wurbs A, Schmidt G, Schröder W, Windhorst W: Risk indication of

genetically modified organisms (GMO). Modelling environmental exposure and dispersal across different scales. Oilseed rape in Northern Germany as an integrated case study. *Ecol Ind* 2009.

29. Spath D, Günther J: [http://www.iao.fraunhofer.de/d/oss_studie.pdf] *webcite Open Source Software - Strukturwandel oder Strohfeuer? - Eine empirische Studie zu Trends und Entwicklungen zum Einsatz von Open Source Software in der öffentlichen Verwaltung und IT-Unternehmen in Deutschland.*
30. GNU General Public License Versions [http://www.opensource.org/licenses/gpl-license.php]
31. Williams S: *Free as in Freedom. Richard Stallman's Crusade for Free Software.* Cambridge: O'Reilly, Sebastopol; 2002.
32. INSPIRE Directive 2007/2/EC [http://inspire.jrc.ec.europa.eu/]
33. Umweltportal Deutschland-PortalU [http://portalu.de]
34. Adams T, Biakowski C, Christl A, Emde A, Thelen B, Trakas A:*Praxishandbuch WebGIS mit Freier Software. Architektur, Beschreibung, Technik und Beispiele mit den Open Source Projekten: UMN MapServer, AVeiN!, PostgreSQL/Post-GIS, Mapbender. CCGIS GbR/terrestris GbR/Geo-Consortium. Bonn.* 2004.
35. GMO location register of the BVL [http://apps2.bvl.bund.de/stareg_web/showflaechen.do]
36. EC: *EU Directive 2004/204/EC: Commission Decision of 23 February 2004 laying down detailed arrangements for the operation of the registers for recording information on genetic modifications in GMOs, provided for in Directive 2001/18/EC of the European Parliament and of the Council (Text with EEA relevance) (notified under document number C(2004) 540).* 2004,L65:20-22.
37. Vaasen A, Gathmann A, Storch J, Bartsch D: Public GMO location register in Germany 2008 - a continuously improved information platform. *Journal of Consumer Protection and Food Safety* 2008, 29-31.
38. Keil M, Kiefl R, Strunz G: *CORINE Land Cover 2000 - Germany. Final report. Project period: 1 May 2001 - 31 December 2004.* German Remote Sensing Data Center Oberpfaffenhofen; 2005.

39. Meynen E, Schmithüsen J, Gellert J, Neef E, Müller-Miny H, Schultze JH: *Handbuch der naturräumlichen Gliederung Deutschlands. Bad Godesberg, 2 Bde*. 1953, 1962.

40. Schröder W, Schmidt G: Defining ecoregions as framework for the assessment of ecological monitoring networks in Germany by means of GIS and classification and regression trees (CART). *Gate to EHS* 2001, 1-9.

41. Agricultural statistics - Statistik lokal 1999, 2003, 2007 [http:/ / www.destatis.de/ jetspeed/ portal/ cms/ Sites/ destatis/ Internet/ DE/ Navigation/ Publikationen/ Querschnittsveroeffentlichungen/ StatistikLokal.psml]

42. Kleppin L, Aden C, Schmidt G, Schröder W: Monitoring of genetically modified maize cultivation by means of WebGIS and Google Maps. In *Geospatial Crossroads @ GI_Forum '08: Proceedings of the Geoinformatics Forum Salzburg*. Edited by Car A, Griesebner G, Strobl J. Heidelberg, Wichmann; 2008:170-179.

43. Spatial information of the federal state of Brandenburg[http:// www.mugv.brandenburg.de/cms/detail.php/bb2.c.515599.de]

44. Monitoring sites within biosphere reserves in Brandenburg [http:// lanuweb.fh-eberswalde.de/oeub/dbf.html]

45. Hofmann F, Epp R, Kalchschmid A, Kratz W, Kruse L, Kuhn U, Maisch B, Müller E, Ober S, Radtke J, Schlechtriemen U, Schmidt G, Schröder W, Von der der Ohe W, Vögel R, Wedl N, Wosniok W: Monitoring of Bt-Maize pollen exposure in the vicinity of the nature reserve Ruhlsdorfer Bruch in northeast Germany 2007 to 2008. *Umweltwiss Schadst Forsch* 2010.

46. Rosi-Marshall EJ, Tank JL, Royer TV, Whiles MR, Evans-White M, Chambers C, Griffith NA, Pokelsek J, Stephen ML: Toxins in transgenic crop byproducts may affect headwater stream. *PNAS* 2007, 104:16204-16208.

47. Schmidt G, Schröder W: GIS-gestützte Analysen zur möglichen Gefährdung von Naturschutzgebieten durch den Anbau gentechnisch veränderter Kulturpflanzen. *Umweltwiss Schadst Forsch* 2009, 21(1):76-93.

48. Mattile HR, Sears MK, Duan JJ: Response of Danaus plexippus to pollen of two new Bt-corn events via laboratory bioassay. *Entomol Exp Appl* 2005, 116:31-41.

49. Graef F, Schmidt G, Schröder W, Graef F, Stachow U: Determinig ecoregions for environmental and GMO monitoring networks. *Environ Monit Assess* 2005, 108:189-203.

50. Finck M, Seitz H, Beismann H: Concepts for General Surveillance: VDI proposals. Standardisation and harmonisation in the field of GMO monitoring. *J Consum Protect Food Safety* 2006, 1:11-14.

51. WebGIS application of the Federal Nature Conservation Agency[http://www.bfn.de/geoinfo/gvo]

52. Risk Register Genetic Engineering Agriculture [http://www.risikoregister.de]

53. Klenke T, Vögele F, Kruse H, Lehmann H: PortalU® & InGrid® - Werkzeuge zur Erstellung, Recherche und Verteilung von Metadaten. In *Angewandte Geoinformatik 2007*. Edited by Strobl J, Blaschke Th, Griesebener G. Heidelberg: Wichmann; 2007:338-343.

54. Rissler J, Mellon M: *International implications of commercialization (Transgenic Crops) in the ecological risks of engineered crops*. Cambridge: MIT Press; 1996.

55. OECD (Organisation for Economic Co-operation and Development): *Emerging Risks in the 21st Century. An Agenda for Action*. Paris: OECD; 2003.

56. Wilkinson MJ, Davenport IJ, Charters YM, Jones AE, Allainguillaume J, Butler HT, Mason DC, Raybould AF: A direct regional scale estimate of transgene movement from genetically modified oilseed rape to its wild progenitors. *Mol Ecol* 2000, 9:983-991.

57. Züghart W, Breckling B: Konzeptionelle Entwicklung eines Monitoring von Umweltwirkungen transgener Kulturpflanzen. *UBA-Texte* 2003, 50/03:1-543.

58. Janicke L, Janicke U: *Enhancement of a diagnostic wind field model for licensing industrial facilities (TA Luft). Umweltforschungsplan des Bundesministeriums für Umwelt, Naturschutz und Reaktorsicherheit. Anlagenbezogener Immissionsschutz. Förderkennzeichen (UFOPLAN) 203 43 256. Im Auftrag des Umweltbundesamtes Berlin*. 2004.

Chapter 9

Effect of Melia Azedarach (Sapindales: Meliaceae) Fruit Extracts on Citrus Leafminer Phyllocnistis Citrella (Lepidoptera: Gracillariidae)

Maher M Mckenna[1], Efat M Abou-Fakhr Hammad[2], and Mohamad T Farran[3]

[1]Department of Plant Protection, Faculty of Agriculture and Veterinary Sciences, Lebanese University, Beirut, Lebanon

[2]Department of Agricultural Sciences, Faculty of Agricultural and Food Sciences, American University of Beirut, Beirut, Lebanon

[3]Department of Animal & Veterinary Sciences, Faculty of Agricultural and Food Sciences, American University of Beirut, Beirut, Lebanon

ABSTRACT

Melia azedarach L. extracts were studied in comparison with selected biorational insecticides against the citrus leafminer *Phyllocnistis citrella* Stainton under field conditions. *Citrus limon* (L.) Burm. F. trees were exposed to: *Melia* extracts of green and mature fruits, Neem oil (30% a.i.), abamectin (1.8% a.i.) and control. Two sprays of each treatment (except for *Melia* mature fruit extract) were executed at 10-d intervals. The live number of the 1st and later (2nd & 3rd) larval instars per leaf were recorded at initial sampling date and at 10-d intervals after each spray application. Results indicated that there were significant differences in the number of live larval instars among treatments. *Melia* extracts and the two biorationals, neem oil and abamectin, decreased the larvae population significantly to lower numbers than that of the control at 10 days after each spray application. However, the decrease caused by neem oil and abamectin was significantly higher than that of *Melia* extracts. Thus, these extracts might be considered as potential alternative with other biorational control methods in management of the leafminer. Further research including bioassays is needed to determine the factors responsible for reducing larvae population and whether these *Melia* extracts can be utilized in future citrus IPM programs as a tool for citrus leafminer management.

INTRODUCTION

The citrus leafminer (CLM) *Phyllocnistis citrella* Stainton (Lepidoptera: Gracillariidae: Phyllocnistinae) is one of the major constraints to citrus production in the world. Effective chemical control of the CLM is limited as the feeding larvae are protected by the leaf cuticle mainly in mines of immature foliage and the pupal stage is also protected by the rolled leaf margins (Jones 2001; Beattie 2004). Several pressures as pesticide resistance has accelerated the search for safe environmentally, toxicologically, selective and efficacious pesticides (Weinzierl and Henn 1991; Dias et al. 2005; Amiri-Besheli 2008, 2009). Biopesticides, including botanicals, can offer a safe and effective alternative to conventional insecticides controlling major insect pests. Botanicals are a promising source of pest control compounds. The pool of plants possessing insecticidal substances is enormous (FAO 1992; Jacobson 1989). These have generated remarkable interest in recent years as potential sources of natural insect control agents. Over 2000 species of plants are known to possess different degrees of insecticidal activity (Jacobson 1975). However, a few plant-based pesticides have already been utilized as plant protection products (Isman 2006). Several secondary plant products represent a large reservoir of chemical structures with biological activity against pests in form of pesticides (Grainage and Ahmed 1988; Schmutterer 1990; Lowery and Isman 1994; Miller and Uetz 1998). Throughout history, plant products have been successfully exploited as insecticides, insect repellents, insect antifeedants (Lydon and Duke 1989) or as altering insect feeding behavior, growth, development, and behavior during mating and oviposition (Yang and Tang 1988).

The Chinaberry tree, *Melia azedarach* L. (Sapindales: Meliaceae), is a deciduous tree often grown for its medicinal uses and for shade or ornament on roadsides. The non-conventional insecticidal effects of extracts or compounds isolated from Meliaceae trees include: partial reduction or complete inhibition of fecundity and egg hatchability, reduction of life span of adults, oviposition deterrence, antifeedant effects, and insect growth regulatory effects at molting (Ascher 1993). *M. azedarach* has been shown to possess many insecticidal effects against different pest species (Shin-Foon 1987; Ascher et al. 1995; Nardo et al. 1997; Valladares et al. 1997; Jazzar and Abou-Fakhr Hammad 2003; Abou-Fakhr Hammad and McAuslane 2006, 2010; Al-Akhras 2010). A limited number of studies have dealt with the use of botanicals against the citrus leafminer. Amiri-Besheli (2011) indicated that the use of biopesticides such as Tondexir extracts from hot pepper, discouraged CLM adults from laying eggs on leaves and posed lower risk to humans and the environment than other pesticides. Pepper extracts worked better on soft bodied insects during the larval stage as the chemical was able to penetrate the leaves. Howard (1993) also found that the 2 biorationals, Avid (abamectin 0.15 EC) and Azatin (60 ppm azadirachtin) with Triton B-1956 spreader-sticker (Rhom & Haas Co., Philadelphia, PA), prevented damage by CLM when sprayed on leaves prior to oviposition by this insect; these sprays were timed just as trees were flushing.

The main strategy in plant protection scopes is to minimize the use of synthetic compounds in benefit of alternatives, such as biorational insecticides. Pesticide resistance and negative effects on non-target organisms are the main problems in pest management (FAO 1992; Franzen 1993). Natural insecticides are relatively harmless against the non-target organisms. In agricultural pest management, botanical insecticides are best suited for use in organic food production in industrialized countries but can play a much greater role in the production and postharvest protection of food in developing countries (Isman 2006). Availability of *Melia* trees allows growers to use available resources to control pests of significant economic importance. The objective of this study is to elucidate the efficacy of *M. azedarach* fruit extracts in management of the citrus leafminer larvae population under field conditions in comparison with other biopesticides.

MATERIALS AND METHODS

The experiment was conducted during the summer of 2011 in an orchard of *Citrus × limon* (L.) Burm.f. Monachello lemon variety (Moussa and El Hajj 2010) located in Tyre, South Lebanon. This variety provides fruits all year round, mainly in winter and spring; the fruit is medium–small, elliptical but tapering at both ends; neck lacking; nipple small and inconspicuous with few or no seeds. Fruit color is yellow at maturity with a thin rind. The tree has nearly thornless branches with dense foliage. During the experimental period (June-July), the maximum ambient temperature ranged between 28 and 32°C whereas the minimum temperature was 20–22°C (http://freemeteo.com/default.asp?pid=15&la=19&cn=LB.). Sampled leaves were examined at the Plant Protection Laboratory, Faculty of Agriculture & Veterinary Sciences, Lebanese University.

Site Description

The lemon orchard is located in Tyrdebba village, 7 kilometers from city of Tyre in South Lebanon, at an altitude of 180 meters. The orchard is adjacent to an olive orchard on the North and to an orange orchard on the South side, while roads bordered East and West sides. The orchard contained 92 lemon trees, distributed over two terraces; one terrace is 2 meters higher than the other and is separated by a steep slope. Trees were 8 years old, distributed in lines, with 3 m spacing between and within those lines. Three years prior to the experiment, trees have been treated once with Supracide (a.i. methidathion 25%) while no other application of pesticides or chemical fertilizers has been given. In order to obtain the essential for the experiment availability of young flush, trees were sufficiently irrigated in late-April 2011.

Preparation of M. Azedarach Extracts

Four Kg of *M. azedarach* green fruits and 2 Kg of *Melia* mature fruits were collected from trees at Tyrdebba village, washed and cleaned with tap water then rinsed with distilled water. Fruits were left to dry at room temperature for 2 days, and then crushed with a hammer into small pieces and ground using a grinding machine (BRAUN Multiquick 3, Hungary). The obtained material was soaked in distilled water, at a rate of 1 g per 5 ml, for 24 hours prior to spraying. The soaked material was filtered using a fine mesh (500 µm) cloth material. About twenty three liters of each of *Melia* green and mature fruits extracts were obtained after filtration.

Experimental Procedure

Four different treatments were tested: *Melia* green fruit extract, *Melia* mature fruit extract, neem oil-30% (Atlantica Agricola, Spain), abamectin 1.8% w/v – Mectin 1.8 EC (Adonis Industrial - S.I.AD s.a.l, Lebanon). Five single-tree replicates per treatment were used, plus 5 trees treated with tap water used as control. Trees were surrounded by untreated border trees, selected so as to avoid spray drifting. All products were applied as foliar sprays to runoff (approximately 4-L), coating upper and lower leaf surfaces, using an 8-L pressure gun sprayer (Zhejiang CNG Plastic Co., China). Also, it's worth mentioning that in windy weather, the canopy of the control trees was covered with plastic sheet to avoid spray drifting from treated trees. Neem oil and abamectin, as two positive controls, were used at the recommended dose of 0.1 µl/100 ml and 0.03 ml/100 ml, respectively. First application was made on June 16th and the second on June 26th. In the second application, 3 treatments plus the control were tested, because there was no sufficient *Melia*mature fruit extract.

Data Collection

Three samplings were made. The first sampling was just before the first spray application, the second sampling was10 days after the first application and before the second spray, and the third sampling was10 days after the second application. In each sampling, 24 young leaves were collected from each tree; 6 from each geographical direction (North, South, East, and West), 3 from the inner and 3 from the outer part of the canopy. The latter factors were selected in order to take representative sample from each tree-replicate. Totally, 600 leaves were collected in each of the first two samplings and 480 in the third. Leaves collected from each direction and tree location were placed in separate labeled small plastic bags and kept transferred within a cooler to the Laboratory, in which they were stored at 10°C until their observation. Numbers of live 1st instar (L1) and 2nd & 3rd instar larvae (L2 & L3) per leaf were recorded. The length of the larva was used for separation of the CLM larval instars; 1st instar larvae were less than 2 mm in length whereas the 2nd and 3rd in stars were between 2 and 3 mm; in addition to their length, the latter instars were translucent and yellowish-green (Kerns et al. 2001).

Statistical Analysis

The experiment was laid out as a three-way factorial arrangement of treatments (5 × 4 × 2) for treatment, geographical direction and leaf location, respectively along with their two- and three-way interactions in a complete randomized design. As there was no significant interaction among the 3 factors, the data were pooled and analyzed as a one-way ANOVA with the treatment as the main factor. Consequently, there were 5 treatments (only 4 treatments in the third sampling) with 5 tree-replicates per treatment. The average numbers of live larvae per leaf were used in the data analysis, after ensuring their normal distribution by transforming the data using sqrt $(x+1)$ with xbeing the number of live larvae per leaf (Gomez and Gomez 1984). The analysis of data was performed using the general linear models procedure, and means were separated by Student Newman Keuls test whenever $P<0.05$ was detected (SAS 1992).

RESULTS

None of the interaction terms among the three main factors: treatment, geographical direction and leaf location was significant. In addition, the factorial analysis showed that direction and location were not significant indicating that CLM had no preference for inner or outside flush of citrus trees nor has preference for infesting one of the directions of the tree. Consequently, data were pooled as indicated above and analyzed with the treatment being the main factor, with mean values presented in Tables 1 & 2. There were no significant differences in the initial number of live larvae of all instars at the first sampling date before applying the 1st spray of each treatment. The initial counts were comparable among treatments with a range of 0.333-0.367 and 0.383-0.492 larva per leaf for the L1 and (L2 & L3) larval instars, respectively.

Table 1: Effects of *Melia azedarach* fruit extracts and selected biorational insecticides on Citrus leafminer first instar larvae (L1) infesting canopy of Lemon trees under field conditions

Average number of live larvae (L1) per leaf*			
Treatments	Before 1st spray**	10 days after 1st spray	10 days after 2nd spray
Control	0.330±0.018	0.517±0.023 a	0.658±0.020 a
Melia green fruit extract	0.330±0.018	0.342±0.018 b	0.400±0.019 b
Melia mature fruit extract	0.350±0.016	0.317±0.019 b	---
Neem oil	0.342±0.017	0.100±0.011 c	0.083±0.010 c
Abamectin	0.367±0.016	0.058±0.009 c	0.067±0.010 c

Measurement values presented are mean+SEM.

*=Means in a column followed by different letters are significantly different (Student Newman Keuls test, $P<0.05$).

**=Initial counts of live larvae were not significantly different at the first date of the experiment before applying the 1st spray for each treatment.

Mckenna *et al.*

Mckenna *et al.* *SpringerPlus* 2013 2:144 doi:10.1186/2193-1801-2-144

Table 2: Effects of *Melia azedarach* fruit extracts and selected biorational insecticides on Citrus leafminer 2nd & 3rd instar larvae (L2& L3) infesting canopy of Lemon trees under field conditions

Average number of live larvae (L2&L3) per leaf*			
Treatments	Before 1st spray**	10 days after 1st spray	10 days after 2nd spray
Control	0.467 ± 0.018	0.567 ± 0.022 a	0.850 ± 0.026 a
Melia green fruit extract	0.492 ± 0.018	0.358 ± 0.017 b	0.417 ± 0.019 b
Melia mature fruit extract	0.492 ± 0.017	0.350 ± 0.017 b	---
Neem oil	0.475 ± 0.018	0.108 ± 0.012 c	0.100 ± 0.011 c
Abamectin	0.383 ± 0.020	0.050 ± 0.010 c	0.033 ± 0.007 c

Measurement values presented are mean + SEM.

* = Means in a column followed by different letters are significantly different (Student Newman Keuls test, $P < 0.05$).

** = Initial counts of live larvae were not significantly different at the first date of the experiment before applying the 1st spray for each treatment.

Mckenna *et al.*

Mckenna *et al.* *SpringerPlus* 2013 **2**:144 doi:10.1186/2193-1801-2-144

There were significant differences in number of L1 among treatments at 10 days after the1st spray application ($F=76.77$; $df=4, 24$; $p<0.0001$) and at 10 days after the 2nd spray application ($F=104.66$; $df=3, 19$; $p<0.0001$). Similarly, there were significant differences ($F=89.02$; $df=4, 24$; $p<0.0001$) and ($F=58.53$; $df=3, 19$; $p<0.0001$) in number of L2 & L3 combined among treatments at the previously mentioned dates of post spray applications, respectively.

There was no significant difference in number of live larvae per leaf between *Melia* green and mature fruit extracts, but both treatments were significantly different from the control. As shown in Table 1, the *Melia* green and mature fruit extracts have caused a significantly low population of L1 (0.342 and 0.317 live larvae per leaf, respectively) compared with the control (0.517) at 10 days after 1st spray application. Also in comparison with the control, the *Melia* green fruit extract resulted in a significantly lower L1 population (0.400 Vs 0.658 live larvae per leaf) at 10 days after the 2nd spray application. Similarly, the *Melia* green and mature fruit extracts have maintained a significantly low population of the L2 & L3 combined (0.358 and 0.350 live larvae per leaf, respectively) in comparison with the control (0.567) at 10 days after the1st spray application (Table 2). In addition, the *Melia* green fruit extract had also maintained a low population of the L2 & L3 (0.417) as compared with 0.850 larvae per leaf for the control at 10 days after the 2nd spray application.

In this study, the *Melia* extracts seem to keep the population of the CLM at a relatively constant level rather than increasing its larval numbers as in the control; the latter showed increased population of the CLM larval number throughout the experiment (Tables 1 & 2). The two commercial insecticides: Neem oil and abamectin had comparative CLM results irrespective of the spray application dates. They had, however, a significantly higher effect than the two *Melia* fruit extracts. In fact, the two insecticides significantly reduced the population of L1 to 0.100 and 0.058 and L2 & L3 (combined) to 0.108 and 0.050 larvae per leaf, respectively at 10 days after the 1stspray application. It is worth mentioning that these insecticides had maintained the larval CLM population at its lowest levels throughout the experimental period (Table 1 & 2).

DISCUSSIONS

Generally, *M. azedarach* fruit extracts exert various effects on leafminer species. The population of the Agromyzid pea leafminer *Liriomyza huidobrensis* (Blanchard) was reduced after surface application of *Melia* fruit extract on swisschard and cucumber plants grown under field and greenhouse conditions, respectively (Abou-Fakhr Hammad et al. 2000a, 2000b). Similarly, *M. azedarach* immature fruit extracts resulted in fewer initiated larval mines of the vegetable leafminer *Liriomyza sativae* Blanchard on cowpea plants (Abou-Fakhr Hammad and McAuslane2010). Furthermore, *Melia* extracts applied on various plants exerted toxic and repellent effect against the nymph and adult whiteflies of *Bemisia* species, respectively (Abou-Fakhr Hammad and McAuslane 2006; Abou-Fakhr Hammad et al. 2001; Abou-Fakhr Hammad et al. 2000a, 2000b).

In the current study, both *Melia* extracts behaved similarly as they yielded comparable results 10 days after the first spray application but unfortunately, the *Melia* mature fruit extract could not be tested thereafter. Extracts of different ages of *Melia* fruits were comparable in their effect against the CLM after the 1st spray application similar to extracts of different ages of *Melia* (Abou-Fakhr Hammad et al. fruits and leaves tested against *B. tabaci*2001). The significantly lower number of CLM live larvae in both fruit extract treatments in comparison with the control (Tables 1& 2), could have been caused by the ingestion of the *Melia* components present in the leaf mines. The fact that the number of larvae remained relatively constant after the second spraying with the green *Melia* extract may indicate a potential toxic or antifeedant effect of this extract. In comparison with the two biorational treatments, however, both *Melia* extracts were less efficient after 10 days of the first application and this efficiency did not improve even after the second spray of the *Melia* green fruit extract. This could be an indication of either low concentrations of the extracts reaching the mines or the result of a weakening effect of the active ingredients of the*Melia* extracts in the plant tissues with time. A weakening effect of azadirachtin (of initially 0.17 g active ingredient/cm trunk diameter) against the litchi stink bug pest, *Tessaratoma papillosa* Drury, in the plant tissues after about 2–3 weeks was reported by Schulte et al. (2006). Furthermore, the persistence of azadirachtin was prolonged in systemic applications compared with its residual life (4–8 days) when foliar sprayed (applied in the current study) under field conditions (Ascher 1993). In addition, the *Melia* extracts used in this study may have lacked the rapid penetration and translaminar action through the leaf tissues; characteristics of neem oil and abamectin, respectively (Mujica et al. 2000).

Different commercial insecticides including botanicals and other biorationals have been tested for the control of CLM. These products differ in their formulation and concentration of the active ingredients. Azadirachtin (as Margson-O ; 25% a.i.), abamectin and mineral oil had a comparable effect against CLM (L1, L2 & L3) on navel oranges under field conditions after 2nd spray application (Abou-Fakhr Hammad and Antar 2003). Howard (1993) reported similar CLM results on Lime seedlings treated with the two biorationals, Avid (abamectin 0.15 EC) and Azatin (60 ppm azadirachtin) with Triton, at the flushing period. Plants sprayed with abamectin, however, remained completely free of leafminer damage, whereas 33 and 50% of azadirachtin treated leaves showed incipient mines 3 and 4 weeks post spraying, respectively; the incipient mines did not progress when observed 5 weeks post spraying with no leaf curling detected. These results indicate that botanical products including *Melia* plant extracts need more time to reveal their insecticidal effects compared with other biorational or synthetic pesticides that are known to have quick knock down action. Furthermore, the extract application efficiency could be enhanced by incorporating mineral oils that improve plant coverage and penetration of chemicals into leaf surfaces (Bográn et al.2006); knowing that *Melia* on tomato plants (Jazzar and Abou-Fakhr Hammad extracts were found to be enhanced by Tween-20 when applied against *B. tabaci*2003).

The compatibility of any potential and emerging pesticide with natural enemies is a basic component in building up Integrated Pest Management (IPM) programs. Villanueva-Jimenez and Hoy (1998) studied the relative toxicity of certain biorational pesticides to the CLM and its parasitoid *Ageniaspis citricola* using several bioassay methods. They classified azadirachtin (Neemix)+oil, and oil alone (FC 435–66) as IPM-compatible insecticides, while azadirachtin (Align)+oil and neem oil (Neemgard) were ranked as semi-compatible insecticides, but avermectin (Agri-Mek)+oil and imidacloprid (Provado) when applied as sprays were considered as IPM-incompatible insecticides. The current results suggest that the *Melia* extract lends itself as a potential plant derived insecticide and merits further investigation in order to classify it as IPM compatible and use it with other control methods. The incorporation of these plant extracts might enhance the effect of natural enemies of the CLM under IPM programs; knowing that several predators and parasitoids (Karamaouna et al. 2010; Kalaitzaki et al. 2011; Tsagkarakis et al. 2013) were found to be important in reducing peak populations of the CLM, but apparently may need to be used in large numbers to provide good control of the pest (Knapp et al.1995; Hoy and Nguyen 2006; Heppner and Fasulo 2010). Limited or no information is available not only on the natural enemies of the CLM in Lebanon, but also on the effect of botanical extracts on these beneficials. *Melia* extracts, however, have been tested against different natural enemy species. Percent parasitism of *Diglyphus isaea* (Walker) (Hymenoptera: Eulophidae), known to lower the populations of the leafminer *L. sativae* (fruit extracts (Abou-Fakhr Hammad and McAuslane Diptera: Agromyzidae) was not affected by *Melia*2010).

The same extracts were also found to have an additive effect on the whitefly *Bemisia argentifolii* Bellows & Perring nymphal mortality when combined with the parasitoid *Eretmocerus rui* Zolnerowich and Rose (Abou-Fakhr Hammad and McAuslane 2006). Furthermore, Melia extracts were found to be harmless to the predator, *Camptotylus reuteri* (Jakovlev), of *B. tabaci* as these extracts did not cause any mortality of the predator (Jazzar and Abou-Fakhr Hammad 2003). Thus, *Melia* extracts deserve further investigations taking into consideration the locally available CLM natural enemies in order to be used in IPM programs.

In conclusion, this work showed that *M. azedarach* extracts clearly had adverse effects on the *P. citrella* by decreasing the number of live CLM larvae (L1 and L2 & L3). These extracts are promising to be used in sustainable agriculture approaches against the CLM and deserve further investigations to fully elucidate their potential use alone and with biological control agents of this pest.

AUTHORS' CONTRIBUTIONS

EMA-FH proposed the topic, specified the required design and chronological steps of the experimental study, followed up on the execution of the experiment, participated in the statistical analysis of the data and adjusted the manuscript draft according to publication needed criteria. MMM carried out the experimental study (field and laboratory work), collected and tabulated the data for the statistical analysis and provided first draft of the manuscript. MTF analyzed the data and helped in their interpretation; also contributed with the corresponding author to the correction of the last revision of the manuscript. All authors read and approved the final manuscript.

ACKNOWLEDGEMENT

The authors would like to express their gratitude to Mr. Abbas Mckenna, for his permission to proceed with this study on his citrus lemon orchard.

REFERENCES

1. Abou-Fakhr Hammad EM, Nemer NM, Hawi ZK, Hanna LT (2000) Responses of the sweetpotato whitefly, *Bemisia tabaci*, to the Chinaberry tree (*Melia azedarach* L.) and its extracts. Ann Appl Biol 137:79-88
2. Abou-Fakhr Hammad EM, Nemer NM, Kawar NS (2000) Efficacy of the Chinaberry tree (Meliaceae) aqueous extracts and certain insecticides against the pea leafminer (Diptera: Agromyzidae). J Agric Sci Vol 134:413-420
3. Abou-Fakhr Hammad EM, Zournajian H, Talhouk S (2001) Efficacy of extracts of *Melia azedarach* L. callus, leaves and fruits against adults of the sweetpotato whitefly *Bemisia tabaci* (Hom. Aleyrodidae). J Appl Entomol 125:483-488
4. Abou-Fakhr Hammad EM, Antar NM (2003) Efficacy of selected insecticides on *Phyllocnistis citrella*Stainton (Lepidoptera: Gracillariidae) in navel orange (*Citrus sinensis* L.). Tests Agrochem Cultivars 24:6-7
5. Abou-Fakhr Hammad E, McAuslane H (2006) Effect of *Melia azedarach* L. Extract on *Bemisia argentifolii* (Homoptera: Aleyrodidae) and its Biocontrol Agent *Eretmocerus rui* (Hymenoptera: Aphelinidae). Environ Entomol 35(3):740-745
6. Abou-Fakhr Hammad E, McAuslane H (2010) Effect of *Melia azedarach* L. Extract on *Liriomyza sativae* (Diptera: Agromyzidae) and its Biocontrol Agent *Diglyphus isaea* (Hymenoptera: Eulophidae). J Food, Agric Environ 8(3&4):1247-1252
7. Al-Akhras HA (2010) Effect of Melia azedarach on Black Aphid Aphis fabae, M.S. Thesis. Beirut, Lebanon: Lebanese University.
8. Amiri-Besheli B (2008) Efficacy of *Bacillus thuringiensis*, mineral oil, insecticidal emulsion and insecticidal gel against *Phyllocnistis citrella* Stainton (Lepidoptera: Gracillariidae). Plant Protec Sci 44(2):68-73
9. Amiri-Besheli B (2009) Toxicity evaluation of Tracer, Palizin, Sirinol, Runner and Tondexir with and without mineral oils on *Phylocnistis citrella* Stainton. Afr J Biotechnol 8(14):3382-3386 http://www.academicjournals.org/ajb/PDF/pdf2009/20Jul/Amiri-Besheli.pdf

10. Amiri-Besheli B (2011) The Effect of Some Botanical Pesticides against Citrus Leafminer (CLM) and Two Spotted Mite (TSM). In: Stoytcheva M (ed) Pesticides in the Modern World - Pesticides Use and Management. InTech.
11. Ascher KRS (1993) Nonconventional insecticidal effects of pesticides available from the neem tree,*Azadirachta indica*. Arch Insect Biochem Physiol 22:433-449
12. Ascher KRS, Schmutterer H, Zebitz CPW, Naqvi SNH (1995) The Persian lilac or Chinaberry tree:*Melia azedarach* L. In: Schmutterer H (ed) The Neem Tree: Source of Unique Natural Products for Integrated Pest Management, Medicine, Weinheim, Germany: Industry and Other Purposes. VCH. pp 605-642
13. Beattie A (2004) Citrus leafminer. Australia: NSW Department of Primary Industries, University of Western Sydney. http://www.dpi.nsw.gov.au/__data/assets/pdf_file/0006/137634/citrus-leafminer.pdf
14. Bográn CE, Ludwig S, Metz B (2006) Using Oils as Pesticides. Texas A&M University System; AgriLife Extension. E-419 11/06. http://insects.tamu.edu/extension/publications/epubs/e-419.cfm
15. Dias C, Carsia P, Simoes N, Oliveira L (2005) Efficacy of *Bacillus thuringiensis* against *Phyllocnistis citrella* (Lepidoptera: Phyllocnitidae). J Econ Entomol 98(6):1880-1883
16. FAO – Food & Agriculture Organization of the United Nations (1992) Pesticide residues in food. Report No. 116. 146
17. Franzen H (1993) Need for development of new strategies for locust control. In: Rembold H (ed) New strategies for locust control 89:9–13, Bonn, Germany: ATSAF.
18. Gomez KA, Gomez AA (1984) Statistical procedure for agricultural research. USA: John Wiley and Sons.
19. Grainage M, Ahmed S (1988) Handbook of plants with pest-control properties. New York: Wiley-Interscience.
20. Heppner JB, Fasulo TR (2010) Citrus Leafminer, *Phyllocnistis citrella Stainton* (Insecta:Lepidoptera: Phyllocnistinae). In: EENY-038 (IN165), one of a series of Featured Creatures from the Entomology and Nematology Department. University of Florida, Florida, USA: Florida Cooperative Extension Service, Institute of Food and Agricultural Sciences. http://edis.ifas.ufl.edu/pdffiles/IN/IN16500.pdf

21. Howard FW (1993) Effectiveness of Two Biorational Substances (Neem and abamectin) against Citrus Leaf Miner. TropicLine 6, No. 6. http://flrec.ifas.ufl.edu/pdfs/TropicLine/TropicLine_06-6a.pdf
22. Hoy MA, Nguyen R (2006) Classical Biological Control of the Citrus Leafminer in Florida. Gainesville. Florida: IPM Florida, Institute of Food and Agricultural Sciences (IFAS) Extension. University of Florida, IFAS, Entomology and Nematology Department, Gainesville, and Division of Plant Industry.http://ipm.ifas.ufl.edu/agriculture/citrus/clmhoy7.shtml
23. Isman MB (2006) Botanical insecticides, deterrents, and repellents in modern agriculture and an increasingly regulated world. Annu Rev Entomol 51:45-66
24. Jacobson M (1975) Insecticides from plants: A review of the literature, 1954–1971. Agricultural Handbook 461. Washington, D.C: U.S Department of Agriculture. p 138
25. Jacobson M (1989) Botanical pesticides, past present and future. In: Arnason JJ, Philogen BR, Morand P (eds) Insecticides of Plant Origin. Proceedings of the American Chemical Society (ACS) Symposium Series, 7th edn. Washington, DC, USA: ACS. pp 1-10
26. Jazzar C, Abou-Fakhr Hammad E (2003) The efficacy of enhanced aqueous extracts of *Melia azedarach* leaves and fruits integrated with the *Camptotylus reuteri* releases against the sweetpotato whitefly nymphs. B Insectol 56(2):269-275
27. Jones J (2001) Citrus leafminer. Arizona: USA: University of Arizona, College of Agriculture & Life Sciences. http://cals.arizona.edu/crops/citrus/insects/leafminer.html
28. Kalaitzaki AP, Tsagkarakis AE, Lykouressis DP (2011) Population fluctuation of *Phyllocnistis citrella*and its parasitoids in two citrus species in Western Crete (Greece). Entomologia Hellenica 20:31-44
29. Karamaouna F, Pascual-Ruiz S, Aguilar-Fenollosa E, Verdú MJ, Urbaneja A, Jacas JA (2010) Changes in predation and parasitism of the citrus leafminer *Phyllocnistis citrella* Stainton (Lepidoptera: Gracillariidae) populations in Spain following establishment of *Citrostichus phyllocnistoides* (Hymenoptera: Eulophidae). Biol Control 52:37-45

30. Kerns D, Wright G, Loghry J (2001) Citrus Leafminer (Phyllocnistis citrella). Cooperative Extension. The University of Arizona. http://ag.arizona.edu/crop/citrus/insects/leafminer.pdf
31. Knapp JL, Albrigo LG, Browning HW, Bullock RC, Heppner JB, Hall DG, Hoy MA, Nyugen R, Peña JE, Stansly PA (1995) Citrus Leafminer, Phyllocnistis citrella Stainton: Current status in Florida-1994. University of Florida, Gainesville, Florida, USA: Institute of Food and Agricultural Sciences, Florida Cooperative Extension Service. p 26
32. Lowery DT, Isman MB (1994) Effects of Neem and azadirachtin on Aphids and their Natural Enemies. In: Hedin PA (ed) Bioregulators for Crop Protection and Pest Control, ACS Symposium Series 557. American Chemical Society, Washington, D.C. USA, Chapter 7:78-91
33. Lydon J, Duke SO (1989) Potential of plants for pesticide use. Herbs, spices and medicinal plants: Recent advances in botany, horticulture, and pharmacology, vol. 4. Phoenix, Arizona, USA: Oryx Press.
34. Miller F, Uetz S (1998) Evaluating Biorational Pesticides for Controlling Arthropod Pests and their Phytotoxic Effects on Greenhouse Crops. HortTechnology 8(2):185-192
35. Moussa Z, El Hajj AK (2010) Citrus production in Lebanon Pre-feasibility Study of IPM of Citrus, Lebanese Agriculture Research Institute (LARI) – Lebanon. Italy: IAM – Bari. pp 1-10
36. Mujica N, Pravatiner M, Cisneros F (2000) Effectiveness of abamectin and plant-oil mixtures on eggs and larvae of the leafminer fly, *Liriomyza huidobrensis* Blanchard. CIP Program Report 1999–2000:161-166
37. Nardo EAB, Costa AS, Lourencao LA (1997) *Melia azedarach* extract as an antifeedant to *Bemisia tabaci* (Homoptera: Aleyrodidae). Fla Entomol 80:92-94
38. SAS Institute (1992) SAS User's Guide: Statistics. Cary, NC, USA: SAS Institute Inc. Schulte MJ, Martin K, Sauerborn J (2006) Effects of azadirachtin injection in litchi trees (*Litchi chinensis* Sonn.) on the litchi stink bug (*Tessaratoma papillosa* Drury) in northern Thailand. J Pest Sci 79:241-250

39. Schmutterer H (1990) Properties and potential of natural pesticides from the neem tree,*Azadirachta indica*. Annu Rev Entomol 35:271-297

40. Shin-Foon C (1987) Experiments on the practical application of chinaberry, *Melia azedarach,* and other naturally occurring insecticides in China. In: Schumutterer H, Ascher KRS (eds) Natural Pesticides from the Neem Tree and Other Tropical Plants, Federal Republic of Germany: GTZ Press, Eschborn. pp 661-668

41. Tsagkarakis AE, Kalaitzaki AP, Lykouressis DP (2013) *Phyllocnistis citrella* Stainton (Lepidoptera: Gracillariidae) and its parasitoids on three citrus varieties in Greece. Phytoparasitica 41(1):23-29

42. Valladares G, Defagó MT, Palacios SM, Carpinella MC (1997) Laboratory evaluation of *Melia azedarach* (Meliaceae) extracts against the elm leaf beetle (Coleoptera: Chrysomelidae). J Econ Entomol 90:747-750

43. Villanueva-Jimenez JA, Hoy MA (1998) Toxicity of pesticides to the citrus leafminer and its parasitoid *Ageniaspis citricola* evaluated to assess their suitability for an IPM program in citrus nurseries. BioControl 43:357-388

44. Weinzierl R, Henn T (1991) Alternatives in Insect management: Biological and Biorational Approaches. North Central Regional Extension, Publication 401. Cooperative Extension Service. Illinois, USA: University of Illinois at Urbana-Champaign. http://web.aces.uiuc.edu/vista/pdf_pubs/altinsec.pdf

45. Yang RZ, Tang CS (1988) Plants used for pest control in China: A literature review. Econ Bot 42:376-406

Citations

CHAPTER 1

Yallappa Rajashekar, Nandagopal Bakthavatsalam, and Thimmappa Shivanandappa, "Botanicals as Grain Protectants," Psyche, vol. 2012, Article ID 646740, 13 pages, 2012. doi:10.1155/2012/646740.

CHAPTER 2

Ipsita Das, Girish Kumar, and Narendra G. Shah, "Microwave Heating as an Alternative Quarantine Method for Disinfestation of Stored Food Grains," International Journal of Food Science, vol. 2013, Article ID 926468, 13 pages, 2013. doi:10.1155/2013/926468.

CHAPTER 3

Y. L. Loko, A. Dansi, M. Tamo, et al., "Storage Insects on Yam Chips and Their Traditional Management in Northern Benin," The Scientific World Journal, vol. 2013, Article ID 484536, 11 pages, 2013. doi:10.1155/2013/484536.

CHAPTER 4

Olufunmilayo E. Ajayi, Arthur G. Appel, and Henry Y. Fadamiro, "Fumigation Toxicity of Essential Oil Monoterpenes to Callosobruchus maculatus (Coleoptera: Chrysomelidae: Bruchinae)," Journal of Insects, vol. 2014, Article ID 917212, 7 pages, 2014. doi:10.1155/2014/917212.

CHAPTER 5

Siriphat Ruengphayak, Ekawat Chaichumpoo, Supaporn Phromphan, Wintai Kamolsukyunyong, Wissarut Sukhaket, Ekapol Phuvanartnarubal, Siripar Korinsak, Siriporn Korinsak, and Apichart Vanavichit, Pseudo-Backcrossing Design for Rapidly Pyramiding Multiple Traits into a Preferential Rice Variety, doi:10.1186/s12284-014-0035-0.

CHAPTER 6

David Rhodes and Michael Stephens, Planted Forest Development in Australia and New Zealand: Comparative Trends and Future Opportunities, doi:10.1186/1179-5395-44-S1-S10.

CHAPTER 7

Winfried Schröder and Gunther Schmidt, Overview of principles and implementations to deal with spatial issues in monitoring environmental effects of genetically modified organisms, doi:10.1186/2190-4715-24-6.

CHAPTER 8

Lukas Kleppin, Gunther Schmidt, and Winfried Schröder, Cultivation of Gmo in Germany: Support of Monitoring and Coexistence Issues by Webgis Technology, doi:10.1186/2190-4715-23-4.

CHAPTER 9

Maher M Mckenna, Efat M Abou-Fakhr Hammad, and Mohamad T Farran, Effect of Melia Azedarach (Sapindales: Meliaceae) Fruit Extracts on Citrus Leafminer Phyllocnistis Citrella (Lepidoptera: Gracillariidae), doi:10.1186/2193-1801-2-144.

Index